VARIATIONAL PRINCIPLES AND METHODS IN THEORETICAL PHYSICS AND CHEMISTRY

This book brings together the essential ideas and methods behind current applications of variational theory in theoretical physics and chemistry. The emphasis is on understanding physical and computational applications of variational methodology rather than on rigorous mathematical formalism.

The text begins with an historical survey of familiar variational principles in classical mechanics and optimization theory, then proceeds to develop the variational principles and formalism behind current computational methodology for bound and continuum quantum states of interacting electrons in atoms, molecules, and condensed matter. It covers multiple scattering theory, as applied to electrons in condensed matter and in large molecules. The specific variational principles developed for electron scattering are then extended to include a detailed presentation of contemporary methodology for electron-impact rotational and vibrational excitation of molecules. The book also provides an introduction to the variational theory of relativistic fields, including a detailed treatment of Lorentz and gauge invariance for the nonabelian gauge field of modern electroweak theory.

Ideal for graduate students and researchers in any field that uses variational methodology, this book is particularly suitable as a backup reference for lecture courses in mathematical methods in physics and theoretical chemistry.

ROBERT K. NESBET obtained his BA in physics from Harvard College in 1951 and his PhD from the University of Cambridge in 1954. He was then a research associate in MIT for two years, before becoming Assistant Professor of Physics at Boston University. He did research at RIAS, Martin Company, Baltimore, the Institut Pasteur, Paris and Brookhaven National Laboratory, before becoming a Staff Member at IBM Almaden Research Center in San Jose in 1962. He acted as an Associate Editor for the *Journal of Computational Physics* and the *Journal of Chemical Physics*, between 1969 and 1974, and was a visiting professor at several universities throughout the world. Professor Nesbet officially retired in 1994, but has continued his research and visiting since then. Over the years he has written more than 270 publications in computational physics, atomic and molecular physics, theoretical chemistry, and solid-state physics.

VARIATIONAL PRINCIPLES AND METHODS IN THEORETICAL PHYSICS AND CHEMISTRY

ROBERT K. NESBET

IBM Almaden Research Center

CAMBRIDGE
UNIVERSITY PRESS

CAMBRIDGE UNIVERSITY PRESS
Cambridge, New York, Melbourne, Madrid, Cape Town, Singapore, São Paulo

Cambridge University Press
The Edinburgh Building, Cambridge CB2 2RU, UK

Published in the United States of America by Cambridge University Press, New York

www.cambridge.org
Information on this title: www.cambridge.org/9780521803915

First published 2003
This digitally printed first paperback version 2005

A catalogue record for this publication is available from the British Library

Library of Congress Cataloguing in Publication data

Nesbet, R. K.
Variational principles and methods in theoretical physics and chemistry/Robert K. Nesbet.
p. cm.
Includes bibliographical references and index.
ISBN 0 521 80391 8
1. Calculus of variations. 2. Mathematical physics. 3. Chemistry, Physical and theoretical–Mathematics. I. Title.
QC20.7.C3 N48 2003
530.15'564–dc21 2002067614

ISBN-13 978-0-521-80391-5 hardback
ISBN-10 0-521-80391-8 hardback

ISBN-13 978-0-521-67575-8 paperback
ISBN-10 0-521-67575-8 paperback

To Anne, Susan, and Barbara

Contents

Preface

As theoretical physics and chemistry have developed since the great quantum revolution of the 1920s, there has been an explosive speciation of subfields, perhaps comparable to the late Precambrian period in biological evolution. The result is that these life-forms not only fail to interbreed, but can fail to find common ground even when placed in proximity on a university campus. And yet, the underlying intellectual DNA remains remarkably similar, in analogy to the findings of recent research in biology. The purpose of this present text is to identify common strands in the substrate of variational theory and to express them in a form that is intelligible to participants in these subfields. The goal is to make hard-won insights from each line of development accessible to others, across the barriers that separate these specialized intellectual niches.

Another great revolution was initiated in the last midcentury, with the introduction of digital computers. In many subfields, there has been a fundamental change in the attitude of practicing theoreticians toward their theory, primarily a change of practical goals. There is no longer a well-defined barrier between theory for the sake of understanding and theory for the sake of predicting quantitative data. Given modern resources of computational power and the coevolving development of efficient algorithms and widely accessible computer program tools, a formal theoretical insight can often be exploited very rapidly, and verified by quantitative implications for experiment. A growing archive records experimental controversies that have been resolved by quantitative computational theory.

It has been said that mathematics is queen of the sciences. The variational branch of mathematics is essential both for understanding and predicting the huge body of observed data in physics and chemistry. Variational principles and methods lie in the bedrock of theory as explanation, and theory as a quantitative computational tool. Quite simply, this is the mathematical foundation of quantum theory, and quantum theory is the foundation of all practical and empirical physics and chemistry, short of a unified theory of gravitation. With this in mind, the present text is subdivided into four parts. The first reviews the variational concepts and formalism that developed over a long history prior to the discovery of quantum mechanics, subdivided into chapters on history, on classical mechanics, and on applied mathematics (severely truncated out of respect for the vast literature already devoted to this subject). The second part covers variational formalism and methodology in subfields concerned with bound states in quantum mechanics. There are separate chapters on time-independent quantum mechanics, on independent-electron models, which may at some point be extended to independent-fermion models as the formalism of the Standard Model evolves, and on time-dependent theory and linear response. The third part develops the variational theory of continuum states, including chapters on multiple scattering theory (the essential formalism for electronic structure calculations in condensed matter), on scattering theory relevant to the true continuum state of a quantum target and an external fermion (with emphasis on methodology for electron scattering by atoms and molecules), continuing to a separate chapter on the currently developing theory of electron-impact rotational and vibrational excitation of molecules. The fourth part develops variational theory relevant to relativistic Lagrangian field theories. The single chapter in this part derives the nonquantized field theory that underlies the quantized theory of the current Standard Model of elementary particles.

This book grew out of review articles in specialized subfields, published by the author over nearly fifty years, including a treatise on variational methods in electron–atom scattering published in 1980. Currently relevant topics have been extracted and brought up to date. References that go more deeply into each of the topics treated here are included in the extensive bibliography. The purpose is to set out the common basis of variational formalism, then to open up channels for further exploration by any reader with specialized interests. The most recent source of this text is a course of lectures given at the Scuola Normale Superiore, Pisa, Italy in 1999. These lectures were presented under the present title, but concentrated on the material in Parts I and II here. The author is indebted to Professor Renato Colle, of Bologna and the Scuola Normale, for making arrangements that made these lectures possible, and to the Scuola Normale Superiore for sponsoring the lecture series.

I

Classical mathematics and physics

This part is concerned with variational theory prior to modern quantum mechanics. The exception, saved for Chapter 10, is electromagnetic theory as formulated by Maxwell, which was relativistic before Einstein, and remains as fundamental as it was a century ago, the first example of a Lorentz and gauge covariant field theory. Chapter 1 is a brief survey of the history of variational principles, from Greek philosophers and a religious faith in God as the perfect engineer to a set of mathematical principles that could solve practical problems of optimization and rationalize the laws of dynamics. Chapter 2 traces these ideas in classical mechanics, while Chapter 3 discusses selected topics in applied mathematics concerned with optimization and stationary principles.

1
History of variational theory

The principal references for this chapter are:

[5] Akhiezer, N.I. (1962). *The Calculus of Variations* (Blaisdell, New York).
[26] Blanchard, P. and Brüning, E. (1992). *Variational Methods in Mathematical Physics* (Springer-Verlag, Berlin).
[78] Dieudonné, J. (1981). *History of Functional Analysis* (North-Holland, Amsterdam).
[147] Goldstine, H.H. (1980). *A History of the Calculus of Variations from the 17th through the 19th Century* (Springer-Verlag, Berlin).
[210] Lanczos, C. (1966). *Variational Principles of Mechanics* (University of Toronto Press, Toronto).
[322] Pars, L.A. (1962). *An Introduction to the Calculus of Variations* (Wiley, New York).
[436] Yourgrau, W. and Mandelstam, S. (1968). *Variational Principles in Dynamics and Quantum Theory*, 3rd edition (Dover, New York).

The idea that laws of nature should satisfy a principle of simplicity goes back at least to the Greek philosophers [436]. The anthropomorphic concept that the engineering skill of a supreme creator should result in rules of least effort or of most efficient use of resources leads directly to principles characterized by mathematical extrema. For example, Aristotle (*De Caelo*) concluded that planetary orbits must be perfect circles, because geometrical perfection is embodied in these curves: "... of lines that return upon themselves the line which bounds the circle is the shortest. That movement is swiftest which follows the shortest line". Hero of Alexandria (*Catoptrics*) proved perhaps the first scientific minimum principle, showing that the path of a reflected ray of light is shortest if the angles of incidence and reflection are equal.

The superiority of circular planetary orbits became almost a religious dogma in the Christian era, intimately tied to the idea of the perfection of God and of His creations. It was replaced by modern celestial mechanics only after centuries in which the concept of esthetic perfection of the universe was gradually superseded by a concept of esthetic perfection of a mathematical theory that could account for the actual behavior of this universe as measured in astronomical observations. Aspects of value-oriented esthetics lay behind Occam's logical "razor" (avoid unnecessary hypotheses), anticipating the later development of observational science and the search for an explanatory theory that was both as general as possible and as simple as possible. The path from Aristotle to Copernicus, Brahe, Kepler, Galileo, and Newton retraces this shift from *a priori* purity of concepts to mathematical theory solidly based on empirical science. The resulting theory of classical mechanics retains extremal principles that are the basis of the variational theory presented here in Chapter 2.

Variational principles have turned out to be of great practical use in modern theory. They often provide a compact and general statement of theory, invariant or covariant under transformations of coordinates or functions, and can be used to formulate internally consistent computational algorithms. Symmetry properties are often most easily derived in a variational formalism.

1.1 The principle of least time

The law of geometrical optics anticipated by Hero of Alexandria was formulated by Fermat (1601–1655) as a principle of least time, consistent with Snell's law of refraction (1621). The time for phase transmission from point P to point Q along a path $\mathbf{x}(t)$ is given by

$$T = \int_P^Q \frac{ds}{v(s)}, \tag{1.1}$$

where ds is a path element, and v is the phase velocity. Fermat's principle is that the value of the integral T should be stationary with respect to any infinitesimal deviation of the path $\mathbf{x}(t)$ from its physical value. This is valid for geometrical optics as a limiting case of wave optics. The mathematical statement is that $\delta T = 0$ for all variations induced by displacements $\delta\mathbf{x}(t)$. In this and subsequent variational formulas, differentials defined by the notation $\delta\cdots$ are small increments evaluated in the limit that quadratic infinitesimals can be neglected. Thus for sufficiently small displacements $\delta\mathbf{x}(t)$, the integral T varies quadratically about its physical value. For planar reflection consider a ray path from $P:(-d,-h)$ to the observation point $Q:(-d,h)$ via an intermediate point $(0,y)$ in the reflection plane $x=0$. Elapsed time in a uniform medium is

$$T(y) = \left\{\sqrt{d^2+(h+y)^2}+\sqrt{d^2+(h-y)^2}\right\}\Big/ v, \tag{1.2}$$

to be minimized with respect to displacements in the reflection plane parametrized by y. The angle of incidence θ_i is defined such that

$$\sin\theta_i = \frac{h+y}{\sqrt{d^2+(h+y)^2}}$$

and the angle of reflection θ_r is defined by

$$\sin\theta_r = \frac{h-y}{\sqrt{d^2+(h-y)^2}}.$$

The law of planar reflection, $\sin\theta_i = \sin\theta_r$, follows immediately from

$$\frac{\partial T}{\partial y} = (\sin\theta_i - \sin\theta_r)/v = 0.$$

To derive Snell's law of refraction, consider the ray path from point $P:(-d,-h)$ to $Q:(d,h)$ via point $(0,y)$ in a plane that separates media of phase velocity $v_i\ (x<0)$ and $v_r\ (x>0)$. The elapsed time is

$$T(y) = v_i^{-1}\sqrt{d^2+(h+y)^2} + v_r^{-1}\sqrt{d^2+(h-y)^2}. \tag{1.3}$$

The variational condition is

$$\frac{\partial T}{\partial y} = \sin\theta_i/v_i - \sin\theta_r/v_r = 0.$$

This determines parameter y such that

$$\frac{\sin\theta_i}{\sin\theta_r} = \frac{v_i}{v_r}, \tag{1.4}$$

giving Snell's law for uniform refractive media.

1.2 The variational calculus

Derivation of a ray path for the geometrical optics of an inhomogeneous medium, given $v(\mathbf{r})$ as a function of position, requires a development of mathematics beyond the calculus of Newton and Leibniz. The elapsed time becomes a functional $T[\mathbf{x}(t)]$ of the path $\mathbf{x}(t)$, which is to be determined so that $\delta T = 0$ for variations $\delta\mathbf{x}(t)$ with fixed end-points: $\delta\mathbf{x}_P = \delta\mathbf{x}_Q = 0$. Problems of this kind are considered in the calculus of variations [5, 322], proposed originally by Johann Bernoulli (1696), and extended to a full mathematical theory by Euler (1744). In its simplest form, the concept of the variation $\delta\mathbf{x}(t)$ reduces to consideration of a modified function $\mathbf{x}_\epsilon(t) = \mathbf{x}(t) + \epsilon\mathbf{w}(t)$ in the limit $\epsilon \to 0$. The function $\mathbf{w}(t)$ must satisfy conditions of continuity that are compatible with those of $\mathbf{x}(t)$. Then $\delta\mathbf{x}(t) = \mathbf{w}(t)\,d\epsilon$ and the variation of the derivative function is $\delta\mathbf{x}'(t) = \mathbf{w}'(t)\,d\epsilon$.

The problem posed by Bernoulli is that of the *brachistochrone*. If two points are connected by a wire whose shape is given by an unknown function $y(x)$ in a vertical plane, what shape function minimizes the time of descent of a bead sliding without friction from the higher to the lower point? The mass of a bead moving under gravity is not relevant. It can easily be verified by trial and error that a straight line does not give the minimum time of passage. Always in such problems, conditions appropriate to physically meaningful solution functions must be specified. Although this is a vital issue in any mathematically rigorous variational calculus, such conditions will be stated as simply as possible here, strongly dependent on each particular application of the theory. Clearly the assumed wire in the brachistochrone problem must have the physical properties of a wire. This requires $y(x)$ to be continuous, but does not exclude a vertical drop. Since no physical wire can have an exact discontinuity of slope, it is reasonable to require velocity of motion along the wire to be conserved at any such discontinuity, so that the hypothetical sliding bead does not come to an abrupt stop or bounce with undetermined loss of momentum. It can easily be verified that a vertical drop followed by a horizontal return to the smooth brachistochrone curve always increases the time of passage. Thus such deviations from continuity of the derivative function do not affect the optimal solution.

The calculus of variations [5, 322] is concerned with problems in which a function is determined by a stationary variational principle. In its simplest form, the problem is to find a function $y(x)$ with specified values at end-points x_0, x_1 such that the integral $J = \int_{x_0}^{x_1} f(x, y, y')dx$ is stationary. The variational solution is derived from

$$\delta J = \int \left\{ \delta y \frac{\partial f}{\partial y} + \delta y' \frac{\partial f}{\partial y'} \right\} dx = 0$$

after integrating by parts to eliminate $\delta y'(x)$. Because

$$\int \delta y' \frac{\partial f}{\partial y'} dx = \delta y \left. \frac{\partial f}{\partial y'} \right|_{x_0}^{x_1} - \int \delta y \frac{d}{dx} \frac{\partial f}{\partial y'} dx,$$

$\delta J = 0$ for fixed end-points $\delta y(x_0) = \delta y(x_1) = 0$ if

$$\frac{\partial f}{\partial y} - \frac{d}{dx} \frac{\partial f}{\partial y'} = 0. \qquad (1.5)$$

This is a simple example of the general form of Euler's equation (1744), derived directly from a variational expression.

Blanchard and Brüning [26] bring the history of the calculus of variations into the twentieth century, as the source of contemporary developments in pure mathematics. A search for existence and uniqueness theorems for variational problems engendered deep studies of the continuity and compactness of mathematical entities that generalize the simple intuitive definitions assumed by Euler and Lagrange. The seemingly self-evident statement that, for free variations of the function $y(x)$,

$$\int \left(\frac{\partial f}{\partial y} - \frac{d}{dx} \frac{\partial f}{\partial y'} \right) \delta y dx = 0$$

implies Euler's equation, was first proven rigorously by Du Bois-Reymond in 1879. With carefully stated conditions on the functions f and y, this made it possible to prove the fundamental theorem of the variational calculus [26], on the existence of extremal solutions of variational problems.

1.2.1 Elementary examples

A geodesic problem requires derivation of the shortest path connecting two points in some system for which distance is defined, subject to constraints that can be either geometrical or physical in nature. The shortest path between two points in a plane follows from this theory. The problem is to minimize

$$J = \int_{x_0}^{x_1} f(x, y, y') dx = \int_{x_0}^{x_1} dx \sqrt{1 + \left(\frac{dy}{dx} \right)^2},$$

where

$$\frac{\partial f}{\partial x} = 0, \quad \frac{\partial f}{\partial y} = 0, \quad \frac{\partial f}{\partial y'} = \frac{y'}{\sqrt{1 + y'^2}}.$$

In this example, Euler's equation takes the form of the geodesic equation

$$\frac{d}{dx} \frac{y'}{\sqrt{1 + y'^2}} = 0.$$

The solution is $y' = const$, or

$$y(x) = y_0 \frac{x_1 - x}{x_1 - x_0} + y_1 \frac{x - x_0}{x_1 - x_0},$$

a straight line through the points x_0, y_0 and x_1, y_1.

In Johann Bernoulli's problem, the brachistochrone, it is required to find the shape of a wire such that a bead slides from point $0, 0$ to x_1, y_1 in the shortest time T under the force of gravity. The energy equation $\frac{1}{2}mv^2 = -mgy$ implies $v = \sqrt{-2gy}$, so that

$$T = \int_0^{x_1} \frac{ds}{v} = \int_0^{x_1} f(y, y') \, dx,$$

where $f(y, y') = \sqrt{-(1 + y'^2)/2gy}$. Because $\partial f/\partial x = 0$, the identity

$$\frac{d}{dx}\left(y'\frac{\partial f}{\partial y'} - f\right) = y'\left(\frac{d}{dx}\frac{\partial f}{\partial y'} - \frac{\partial f}{\partial y}\right),$$

and the Euler equation imply an integral of motion,

$$y'\frac{\partial f}{\partial y'} - f = \frac{-1}{\sqrt{-2gy(1 + y'^2)}} = const.$$

On combining constants into the single parameter a this implies

$$1 + \left(\frac{dy}{dx}\right)^2 = \frac{-2a}{y}.$$

The solution for a bead starting from rest at the coordinate origin is a cycloid, determined by the parametric equations $x = a(\phi - \sin\phi)$ and $y = a(\cos\phi - 1)$. This curve is generated by a point on the perimeter of a circle of radius a that rolls below the x-axis without slipping. The lowest point occurs for $\phi = \pi$, with $x_1 = \pi a$ and $y_1 = -2a$. By adding a constant ϕ_0 to ϕ, a can be adjusted so that the curve passes through given points x_0, y_0 and x_1, y_1.

1.3 The principle of least action

Variational principles for classical mechanics originated in modern times with the principle of least action, formulated first imprecisely by Maupertuis and then as an example of the new calculus of variations by Euler (1744) [436]. Although not stated explicitly by either Maupertuis or Euler, stationary action is valid only for motion in which energy is conserved. With this proviso, in modern notation for generalized coordinates,

$$\delta \int_P^Q \mathbf{p} \cdot d\mathbf{q} = 0, \tag{1.6}$$

for a path from system point P to system point Q.

For a particle of mass m moving in the (x, y) plane with force per mass (X, Y), instantaneous motion is described by velocity v along the trajectory. An instantaneous radius of curvature ρ is defined by angular momentum $\ell = mv\rho$ such that the centrifugal force mv^2/ρ balances the true force normal to the trajectory. Hence, following Euler's derivation, Newtonian mechanics implies that

$$\frac{v^2}{\rho} = \frac{Y dx - X dy}{\sqrt{dx^2 + dy^2}}$$

along the trajectory. The principle of least action requires the action integral per unit mass

$$\int v\,ds = \int v\,dx\sqrt{1+\left(\frac{dy}{dx}\right)^2}$$

to be stationary. The variation of v along the trajectory is determined for fixed energy $E = T + V$ by

$$v\,dv = -\frac{1}{m}\left(\frac{\partial V}{\partial x}dx + \frac{\partial V}{\partial y}dy\right) = X dx + Y dy.$$

Thus $v\frac{\partial v}{\partial x} = X$ and $v\frac{\partial v}{\partial y} = Y$. Euler's equation then takes the form

$$\frac{d}{dx}\left(\frac{vy'}{\sqrt{1+y'^2}}\right) - \frac{Y}{v}\sqrt{1+y'^2} = 0,$$

where $y' = dy/dx$. The local curvature of a trajectory is defined by

$$\frac{1}{\rho} = \frac{d}{dx}[y'/(1+y'^2)^{\frac{1}{2}}] = y''/(1+y'^2)^{\frac{3}{2}}.$$

Using this formula and $\frac{dv}{dx} = \frac{X+Yy'}{v}$, Euler's equation implies

$$\frac{v}{\rho} + \frac{(X+Yy')y'}{v\sqrt{1+y'^2}} - \frac{Y}{v}\sqrt{1+y'^2} = 0.$$

This reproduces the formula derived directly from Newtonian mechanics:

$$\frac{v^2}{\rho} = \frac{Y - Xy'}{\sqrt{1+y'^2}} = \frac{Y dx - X dy}{\sqrt{dx^2+dy^2}}.$$

Euler's proof of the least action principle for a single particle (mass point in motion) was extended by Lagrange (*c.* 1760) to the general case of mutually interacting particles, appropriate to celestial mechanics. In Lagrange's derivation [436], action along a system path from initial coordinates P to final coordinates Q is defined by

$$A = \sum_a m_a \int_P^Q v_a\,ds_a = \sum_a m_a \int_P^Q \dot{\mathbf{x}}_a \cdot d\mathbf{x}_a. \tag{1.7}$$

Variations about a true dynamical path are defined by coordinate displacements $\delta\mathbf{x}_a$. Velocity displacements $\delta\dot{\mathbf{x}}_a$ are constrained so as to maintain invariant total energy. This implies modified time values at the displaced points [146]. The energy constraint condition is

$$\delta E = \sum_a \left(m_a\dot{\mathbf{x}}_a \cdot \delta\dot{\mathbf{x}}_a + \frac{\partial V}{\partial \mathbf{x}_a}\cdot\delta\mathbf{x}_a\right) = 0.$$

The induced variation of action is

$$\delta A = \sum_a m_a \int_P^Q (\dot{\mathbf{x}}_a \cdot d\delta\mathbf{x}_a + \delta\dot{\mathbf{x}}_a \cdot d\mathbf{x}_a)$$
$$= \sum_a m_a \dot{\mathbf{x}}_a \cdot \delta\mathbf{x}_a|_P^Q - \sum_a m_a \int_P^Q (d\dot{\mathbf{x}}_a \cdot \delta\mathbf{x}_a - \dot{\mathbf{x}}_a dt \cdot \delta\dot{\mathbf{x}}_a),$$

on integrating by parts and using $d\mathbf{x}_a = \dot{\mathbf{x}}_a dt$. The final term here can be replaced, using the energy constraint condition. Then, using $d\dot{\mathbf{x}}_a = \ddot{\mathbf{x}}_a dt$,

$$\delta A = \sum_a m_a \dot{\mathbf{x}}_a \cdot \delta\mathbf{x}_a|_P^Q - \sum_a \int_P^Q \left(m_a \ddot{\mathbf{x}}_a + \frac{\partial V}{\partial \mathbf{x}_a} \right) \cdot \delta\mathbf{x}_a dt.$$

If the end-points are fixed, the integrated term vanishes, and A is stationary if and only if the final integral vanishes. Since $\delta\mathbf{x}_a$ is arbitrary, the integrand must vanish, which is Newton's law of motion. Hence Lagrange's derivation proves that the principle of least action is equivalent to Newtonian mechanics if energy is conserved and end-point coordinates are specified.

2
Classical mechanics

The principal references for this chapter are:

[26] Blanchard, P. and Brüning, E. (1992). *Variational Methods in Mathematical Physics: A Unified Approach* (Springer-Verlag, Berlin).
[146] Goldstein, H. (1983). *Classical Mechanics*, 2nd edition (Wiley, New York).
[187] Jeffreys, H. and Jeffreys, B.S. (1956). *Methods of Mathematical Physics*, 3rd edition (Cambridge University Press, New York).
[208] Kuperschmidt, B.A. (1990). *The Variational Principles of Dynamics* (World Scientific, New York).
[210] Lanczos, C. (1966). *Variational Principles of Mechanics* (University of Toronto Press, Toronto).
[240] Mercier, A. (1959). *Analytical and Canonical Formalism in Physics* (Interscience, New York).
[323] Pauli, W. (1958). *Theory of Relativity*, tr. G. Field (Pergamon Press, New York).
[393] Synge, J.L. (1956). *Relativity: the Special Theory* (Interscience, New York).
[436] Yourgrau, W. and Mandelstam, S. (1968). *Variational Principles in Dynamics and Quantum Theory*, 3rd edition (Dover, New York).

2.1 Lagrangian formalism

Newton's equations of motion, stated as "*force equals mass times acceleration*", are strictly true only for mass points in Cartesian coordinates. Many problems of classical mechanics, such as the rotation of a solid, cannot easily be described in such terms. Lagrange extended Newtonian mechanics to an essentially complete nonrelativistic theory by introducing generalized coordinates q and generalized forces Q such that the work done in a dynamical process is $\sum_k Q_k dq_k$ [436]. Since this must be the same when expressed in Cartesian coordinates, it follows that $Q_k = \sum_a \mathbf{X}_a \cdot \frac{\partial \mathbf{X}_a}{\partial q_k}$, where the Newtonian force is $\mathbf{X}_a = -\frac{\partial V}{\partial \mathbf{X}_a}$. Equivalently, if the potential function is $V(\{q\})$ in generalized coordinates, then $Q_k = -\frac{\partial V}{\partial q_k}$. The Newtonian kinetic energy $T = \frac{1}{2}\sum_a m_a \dot{\mathbf{x}}_a^2$ defines momenta $\mathbf{p}_a = \frac{\partial T}{\partial \dot{\mathbf{x}}_a} = m_a \dot{\mathbf{x}}_a$, which becomes $p_k = \frac{\partial T}{\partial \dot{q}_k}$ when kinetic energy is expressed as $T(\{q, \dot{q}\})$. The equations of motion $\dot{\mathbf{p}}_a = \mathbf{X}$ transform into $\dot{p}_k = Q_k$. Although this can be shown by direct transformation, a more elegant and ultimately more general procedure is to prove the variational principle of Hamilton in a form that is valid for any choice of generalized coordinates.

2.1.1 Hamilton's variational principle

The principle of least action suffers from the awkward constraint that energy must be fixed on nonphysical displaced trajectories. Following the introduction by Lagrange of a dynamical formalism using generalized coordinates, it was shown by Hamilton that a revised variational principle could be based on a new definition of action that had the full generality of Lagrange's theory. Hamilton's definition of the action integral is

$$I = \int_{t_0}^{t_1} L\, dt, \tag{2.1}$$

where the Lagrangian L is an explicit function of generalized coordinates $\{q\}$ and their time derivatives $\{\dot{q}\}$,

$$L = T(q, \dot{q}) - V(q, \dot{q}). \tag{2.2}$$

Generalized momenta are defined by $p_k = \frac{\partial L}{\partial \dot{q}_k}$ and generalized forces are defined by $Q_k = \frac{\partial L}{\partial q_k}$. When applied using t as the independent variable, Euler's variational equation for the action integral I takes the form of Lagrange's equations of motion

$$\frac{d}{dt}\frac{\partial L}{\partial \dot{q}} - \frac{\partial L}{\partial q} = 0. \tag{2.3}$$

Hamilton considered variations $\delta q(t)$ that define continuous generalized trajectories with fixed end-points, $q(t_0) = q_0$ and $q(t_1) = q_1$. The variational expression is

$$\delta I = \int_{t_0}^{t_1} dt \left\{ \delta \dot{q}(t) \frac{\partial L}{\partial \dot{q}} + \delta q(t) \frac{\partial L}{\partial q} \right\} = 0.$$

Following the logic of Euler, after integrating by parts to replace the term in $\delta \dot{q}$ by one in δq, this implies the Lagrangian equations of motion.

2.1.2 Dissipative forces

Hamilton's principle exploits the power of generalized coordinates in problems with static or dynamical constraints. Going beyond the principle of least action, it can also treat dissipative forces, not being restricted to conservative systems. If energy loss during system motion due to a nonconservative force is $-dW$, where $dW = \sum_k \hat{Q}_k dq_k$, then a dissipative action function is defined by its variation $\delta I = \int_{t_0}^{t_1} (\delta L + \delta W) dt$. Hamilton's variational principle implies modified equations of motion

$$\frac{d}{dt}\frac{\partial L}{\partial \dot{q}_k} - \frac{\partial L}{\partial q_k} = \hat{Q}_k.$$

The dissipative term takes the form of a nonconservative force, $\hat{Q}_k$, acting on the generalized coordinate q_k.

2.1.3 Lagrange multiplier method for constraints

The differential form $dX_j = \sum_k a_{kj} dq_k + a_{kt} dt = 0$ defines a general linear constraint condition. For an integrable or *holonomic* constraint, expressed by $X_j(\{q_k\}, t) = 0$, the coefficients are partial derivatives of X_j such that $a_{kj} = \frac{\partial X_j}{\partial q_k}$ and $a_{kt} = \frac{\partial X_j}{\partial t}$. Nonintegrable or *nonholonomic* constraints are defined by the differential form. In applications of Hamilton's principle, δX_j differs from dX_j because the displacement of each point on a trajectory is defined such that $\delta t = 0$. Hence the coefficients a_{kt} drop out of δX_j.

Lagrange introduced the very powerful method of Lagrange multipliers to incorporate such constraints into the formalism of analytical dynamics. The basic idea is that if a term $X_j \lambda_j$ is added to L, its gradients provide a generalized internal force that dynamically enforces the desired constraint. The Lagrange multiplier λ_j provides a generalized coordinate whose value is to be determined so that the constraint condition is satisfied. This artificial coordinate characterizes an effective potential well whose extremum stabilizes a system conformation that satisfies the constraint condition. Because the added term $X_j \lambda_j$ vanishes when the constraint condition is satisfied, it does not change any physical properties of the system except by avoiding dynamical states incompatible with the constraint.

The equations of motion, generalized to include holonomic or nonholonomic constraints with Lagrange multipliers, are

$$\frac{d}{dt}\frac{\partial L}{\partial \dot{q}_k} - \frac{\partial L}{\partial q_k} - \sum_j a_{kj} \lambda_j = 0.$$

These equations are to be supplemented by the alternative constraint conditions:

$$X_j(\{q_k\}, t) = 0,$$

$$\sum_k a_{kj} \dot{q}_k + a_{kt} = 0,$$

for holonomic and nonholonomic constraints, respectively. The number of equations equals the number of unknown quantities $\{q_k, \lambda_j\}$ in either case. Forces of constraint $\hat{Q}_k = \sum_j a_{kj}\lambda_j$ appear in the modified Lagrange equations.

A simple example of the use of Lagrange multipliers is a hoop of mass M and radius r rolling without slipping down an inclined plane [146]. Appropriate generalized coordinates are x, the distance from the top of the plane to the point of contact and θ, the angle of rotation of the hoop. If the plane is at a fixed angle ϕ from the horizontal, kinetic and potential energy are, respectively, $T = \frac{1}{2}M\dot{x}^2 + \frac{1}{2}Mr^2\dot{\theta}^2$, $V = -Mgx\sin\phi$ subject to the differential constraint $dX = rd\theta - dx = 0$. $\delta(T - V + X\lambda)$ replaces δL, where λ is the Lagrange multiplier for the constraint. The coefficients in the differential constraint form are $a_\theta = r$ and $a_x = -1$. The equations of motion and constraint are

$$\begin{aligned} M\ddot{x} - Mg\sin\phi + \lambda &= 0, \\ Mr^2\ddot{\theta} - \lambda r &= 0, \\ r\dot{\theta} - \dot{x} &= 0. \end{aligned}$$

The solution is simplified because $r\ddot{\theta} = \ddot{x}$ implies $M\ddot{x} = \lambda$, so that $\ddot{x} = g\sin\phi/2$. Hence $\lambda = Mg\sin\phi/2$, and $-\lambda$ is the translational force of constraint. The angular acceleration is $\ddot{\theta} = g\sin\phi/2r$. The hoop rolls with half the unconstrained translational acceleration.

2.2 Hamiltonian formalism

Generalizing Newton's equations of motion, Lagrange's equations also set the time rate of change of momenta p equal to forces Q. Hamilton recognized that these generalized momenta could replace the time derivatives $\dot{q}$ as fundamental variables of the theory. This is most directly accomplished by a Legendre transformation, as described in the following subsection.

2.2.1 The Legendre transformation

Given $F(x, y)$ such that

$$dF = u(x, y)\,dx + v(x, y)\,dy = \frac{\partial F}{\partial x}dx + \frac{\partial F}{\partial y}dy,$$

it is often desirable to transform to different independent variables (u, y). The alternative function $G(u, y) = F(x, y) - ux$ characterizes a Legendre transformation, such that $dG = dF - u\,dx - x\,du = v\,dy - x\,du$. Thus the partial derivatives are

$$\frac{\partial G}{\partial y} = v(x, y), \qquad \frac{\partial G}{\partial u} = -x,$$

expressed as functions of the original variables.

This transformation is used to define thermodynamic functions that depend on easily measurable variables such as pressure and temperature [57, 403]. For example, the thermodynamic function X:enthalpy is determined by V:volume, S:entropy, T:temperature, and P:pressure through the differential form $dX = T\,dS + V\,dP$. A Legendre transformation from the abstractly defined entropy S to the directly measurable temperature T defines the Gibbs function $G = X - TS$. Changes $dG = -S\,dT + V\,dP$ of the Gibbs function are determined by temperature and pressure changes. In an isothermal process, $\frac{\partial G}{\partial P} = V$ is the system volume. In an isobaric process, $\frac{\partial G}{\partial T} = -S$ directly measures the entropy.

2.2.2 Transformation from Lagrangian to Hamiltonian

Independent variables $\{\dot{q}\}$ in a general time-dependent Lagrangian L are replaced by momenta $\{p\}$ using a Legendre transformation that defines the Hamiltonian function $H(p, q, t) = \sum_k p_k\dot{q}_k - L(q, \dot{q}, t)$. Using the definition $p_k = \frac{\partial L}{\partial \dot{q}_k}$ and the Lagrangian equations of motion, the transformation removes terms in $d\dot{q}_k$ from the differential form $dH = \sum_k\{\dot{q}_k dp_k - \dot{p}_k dq_k\} - \frac{\partial L}{\partial t}dt$. The implied partial derivatives give Hamilton's equations of motion

$$\begin{aligned} \dot{q}_k &= \frac{\partial H}{\partial p_k}, \\ -\dot{p}_k &= \frac{\partial H}{\partial q_k}. \end{aligned} \tag{2.4}$$

Alternatively, an auxiliary variable w_k can be introduced, constrained to be dynamically equal to $\dot{q}_k$ using a Lagrange multiplier that turns out to be the substituted variable p_k. The constraint condition in this ingenious procedure is $X_k = \dot{q}_k - w_k = 0$. The modified Lagrangian is

$$\begin{aligned} L' &= L(q, w) + \sum_k p_k(\dot{q}_k - w_k) \\ &= \sum_k p_k\dot{q}_k - \left[\sum_k p_k w_k - L(q, w)\right] \\ &= \sum_k p_k\dot{q}_k - H(p, q, w), \end{aligned}$$

in agreement with the Legendre transformation. The constrained equations of motion are

$$\frac{\partial L'}{\partial p_k} = \dot{q}_k - w_k = \dot{q}_k - \frac{\partial H}{\partial p_k} = 0;$$
$$\frac{d}{dt}\frac{\partial L'}{\partial \dot{q}_k} - \frac{\partial L'}{\partial q_k} = \dot{p}_k + \frac{\partial H}{\partial q_k} = 0;$$
$$\frac{\partial L'}{\partial w_k} = \frac{\partial L}{\partial w_k} - p_k = -\frac{\partial H}{\partial w_k} = 0.$$

The first and second of these equations are Hamilton's equations of motion. The first and third establish $w_k = \dot{q}_k$ and $p_k = \frac{\partial L}{\partial w_k}$ as dynamical conditions, equivalent to $p_k = \frac{\partial L}{\partial \dot{q}_k}$.

2.2.3 Example: the central force problem

In spherical polar coordinates, for one mass point moving in a central potential $V = V(r)$,

$$T = \frac{1}{2}m(\dot{r}^2 + r^2\dot{\theta}^2).$$

The Hamiltonian is $H = \frac{p_r^2}{2m} + \frac{p_\theta^2}{2mr^2} + V(r)$, where $p_r = \frac{\partial T}{\partial \dot{r}} = m\dot{r}$, and $p_\theta = \frac{\partial T}{\partial \dot{\theta}} = mr^2\dot{\theta}$. Using the relevant partial derivatives, Hamilton's equations of motion are

$$\dot{p}_\theta = 0; \qquad \dot{\theta} = \frac{p_\theta}{mr^2};$$
$$\dot{p}_r = mr\dot{\theta}^2 - \frac{\partial V}{\partial r}; \qquad \dot{r} = \frac{p_r}{m}.$$

For specified energy E and angular momentum $p_\theta = const = \ell$, on solving $H = E$ for p_r,

$$m\dot{r} = p_r = [2m(E - V(r)) - \ell^2/r^2]^{\frac{1}{2}}.$$

A trajectory $r(t)$ is obtained by integrating $\dot{r}$ between turning points, where the argument of the square root vanishes. Using $d\theta = \frac{\ell}{mr^2}\frac{dr}{\dot{r}}$, this formula for $\dot{r}$ provides an equation to be integrated for a classical orbit,

$$\frac{d\theta}{dr} = \frac{\ell}{r^2}[2m(E - V(r)) - \ell^2/r^2]^{-\frac{1}{2}}.$$

2.3 Conservation laws

Conservation of energy follows directly from Hamilton's equations. The differential change of the Hamiltonian along a trajectory is

$$dH = \sum_k (\dot{q}_k dp_k - \dot{p}_k dq_k) + \frac{\partial H}{\partial t} dt,$$

using the Hamiltonian equations of motion. The total time derivative is

$$\frac{dH}{dt} = \sum_i (\dot{q}_k \dot{p}_k - \dot{p}_k \dot{q}_k) + \frac{\partial H}{\partial t} = \frac{\partial H}{\partial t}.$$

Hence, stated as a theorem,

$$\frac{\partial H}{\partial t} = 0 \Rightarrow H(t) = E = const,$$

and energy is conserved unless H is explicitly time-dependent.

Since Hamilton's equations imply that $\dot{p}_k = 0$ if $\frac{\partial H}{\partial q_k} = 0$, p_k is a constant of motion if q_k is such an *ignorable* coordinate. An ingenious choice of generalized coordinates can produce such constants and simplify the numerical or analytic task of integrating the equations of motion.

The close connection between symmetry transformations and conservation laws was first noted by Jacobi, and later formulated as Noether's theorem: invariance of the Lagrangian under a one-parameter transformation implies the existence of a conserved quantity associated with the generator of the transformation [304]. The equations of motion imply that the time derivative of any function $\Xi(p, q)$ is

$$\dot{\Xi} = \sum_k \left(\frac{\partial \Xi}{\partial q_k} \dot{q}_k + \frac{\partial \Xi}{\partial p_k} \dot{p}_k \right) = \sum_k \left(\frac{\partial \Xi}{\partial q_k} \frac{\partial H}{\partial p_k} - \frac{\partial \Xi}{\partial p_k} \frac{\partial H}{\partial q_k} \right),$$

which defines a classical Poisson bracket $\{\Xi, H\}$. A constant of motion is characterized by $\{\Xi, H\} = 0$. Invariance under a symmetry transformation implies such a vanishing Poisson bracket for the symmetry generator.

By Noether's theorem, invariance of the Lagrangian under an infinitesimal time displacement implies conservation of energy. This is consistent with the direct proof of energy conservation given above, when L and by implication H have no explicit time dependence. Define a continuous time displacement by the transformation $t = t' + \alpha(t')$ where $\alpha(t_0) = \alpha(t_1) = 0$, subject to $\alpha \to 0$. Time intervals on the original and displaced trajectories are related by $dt = (1 + \alpha')dt'$ or $dt' = (1 - \alpha')dt$. The transformed Lagrangian is

$$L(q, \dot{q}) = L(q, q'(1 - \alpha')) = L(q, q') - \sum_k \frac{\partial L}{\partial \dot{q}_k} \dot{q}_k \alpha'.$$

The transformed action integral is

$$I_\alpha = \int_{t_0}^{t_1} L(q, \dot{q})\, dt = \int_{t_0}^{t_1} L(q, q'(1-\alpha'))(1+\alpha')\, dt'.$$

In the limit $\alpha \to 0$ this is

$$I_{\alpha \to 0} = I - \int_{t_0}^{t_1} \left(\sum_k \frac{\partial L}{\partial \dot{q}_k} \dot{q}_k - L \right) \alpha' dt'.$$

Treating α as a generalized coordinate, its equation of motion is

$$\frac{d}{dt'} \frac{\partial L'}{\partial \alpha'} = -\frac{d}{dt'} \left(\sum_k p_k \dot{q}_k - L \right) = 0,$$

so that $H = \sum_k p_k \dot{q}_k - L = const = E$.

Similarly for translational invariance, an infinitesimal coordinate translation is defined for a single particle by

$$\mathbf{x}'_i = \mathbf{x}_i + \boldsymbol{\alpha}(t), \qquad \boldsymbol{\alpha} \to \mathbf{0}.$$

Assuming invariant potential energy, the transformed kinetic energy is

$$T_\alpha = \frac{1}{2} \sum_i m_i (\dot{\mathbf{x}}_i + \dot{\boldsymbol{\alpha}})^2 = T + \sum_i m_i \dot{\mathbf{x}}_i \cdot \dot{\boldsymbol{\alpha}} + O(\alpha^2).$$

Treating the components of $\boldsymbol{\alpha}$ as generalized coordinates, an equation of motion is implied in the form

$$\frac{\partial L_\alpha}{\partial \dot{\boldsymbol{\alpha}}} = \sum_i m_i \dot{\mathbf{x}}_l = const = \mathbf{P},$$

the conservation law for linear momentum.

2.4 Jacobi's principle

As an introduction to relativistic dynamics, it is of interest to treat time as a dynamical variable rather than as a special system parameter distinct from particle coordinates. Introducing a generic global parameter τ that increases along any generalized system trajectory, the function $t(\tau)$ becomes a dynamical variable. In special relativity, this immediately generalizes to $t_i(\tau)$ for each independent particle, associated with spatial coordinates $\mathbf{x}_i(\tau)$. Hamilton's action integral becomes

$$I = \int L t' d\tau, \tag{2.5}$$

where $t' = dt/d\tau$. Limiting the discussion to conservative systems, with fixed energy E, the modified Lagrangian Lt' does not depend on t. Hence the generalized momentum $p_t = \frac{\partial}{\partial t'}(Lt')$ is a constant of motion. In detail,

$$p_t = L + \sum_i \frac{\partial L}{\partial \dot{q}_i} t' \frac{\partial}{\partial t'}\left(\frac{q_i'}{t'}\right) = L - \sum_i p_i \dot{q}_i = -H = -E.$$

Again anticipating relativistic dynamics, energy is related to momenta as time is related to spatial coordinates.

Since time here is an "ignorable" variable, it can be eliminated from the dynamics by subtracting $p_t t'$ from the modified Lagrangian and by solving $H = E$ for t' as a function of the spatial coordinates and momenta. This produces Jacobi's version of the principle of least action as a dynamical theory of trajectories, from which time dependence has been removed. The modified Lagrangian is

$$\Lambda = Lt' - p_t t' = (T - V + E)t' = 2Tt',$$

such that

$$\frac{\partial}{\partial t'}\Lambda = \frac{\partial}{\partial t'}(Lt') - p_t = 0.$$

Since kinetic energy is positive, the action integral $A = \int 2Tt'd\tau$ is nondecreasing. This suggests using a global parameter $\tau = s$ defined by the Riemannian line element

$$ds^2 = \sum_{i,j} m_{ij}\, dq_i\, dq_j,$$

where the positive-definite matrix m_{ij} defines a mass tensor as a function of generalized coordinates. Then

$$\dot{s}^2 = \sum_{i,j} m_{ij}\dot{q}_i\dot{q}_j = 2T.$$

This makes it possible to express t' as a function of the generalized coordinates and momenta. $\dot{s}^2 = (s'/t')^2 = 2T$ implies that $ds = (2T)^{\frac{1}{2}}dt$, or $t' = (2T)^{-\frac{1}{2}}s'$. The reduced action integral, originally derived by Jacobi, is

$$A = \int 2Tt'd\tau = \int (2T)^{\frac{1}{2}}s'd\tau = \int (2T)^{\frac{1}{2}}ds.$$

Because variations are restricted to those that conserve energy, $2T = 2(E - V)$ and $ds = (2(E - V))^{\frac{1}{2}}dt$ along the varied trajectories. The action integral becomes

$$A = \int 2(E - V)dt = \int \sum_{i,j} m_{ij}\dot{q}_i\dot{q}_j\, dt - \int \sum_i p_i\, dq_i.$$

$\delta A = 0$ subject to the energy constraint restates the principle of least action. When the external potential function is constant, the definition of ds as a path element implies that the system trajectory is a geodesic in the Riemann space defined by the mass tensor m_{ij}. This anticipates the profound geometrization of dynamics introduced by Einstein in the general theory of relativity.

2.5 Special relativity

The concept of a mass point remains valid, but a time interval dt can no longer be treated as a nondynamical parameter. Einstein's basic postulate [323, 393] is that the interval ds between two space-time events is characterized by the invariant expression

$$ds^2 = c^2dt^2 - dx^2 - dy^2 - dz^2,$$

where c is the constant speed of light. The space and time intervals can be measured in any reference frame in uniform linear motion, referred to as an inertial frame. Space and time intervals measured in two such inertial frames are related by a linear Lorentz transformation, parametrized by the velocity and direction of relative motion. It is convenient to introduce Minkowski space-time coordinates

$$dx_\mu = (dx_1, dx_2, dx_3, dx_4) = (dx, dy, dz, ic\,dt),$$

such that $ds^2 = -dx_\mu dx_\mu$, using the summation convention of Einstein. A covariant 4-vector A_μ is defined by four quantities that are related by the appropriate Lorentz transformation when measured in two different inertial frames.

Since dt cannot be singled out for special treatment, the covariant generalization of Hamilton's variational principle for a single particle requires an invariant action integral

$$I = \int \Lambda(x_\mu, u_\mu, \tau)\,d\tau,$$

where $c^2d\tau^2 = ds^2 > 0$ for a *timelike* interval defines an invariant *proper time* interval $d\tau$. $u_\mu = dx_\mu/d\tau$ defines a covariant velocity 4-vector. The variational condition $\delta I = 0$, for fixed end-points x^0_μ and x^1_μ separated by a timelike interval, implies relativistic Lagrangian equations of motion

$$\frac{d}{d\tau}\frac{\partial \Lambda}{\partial u_\mu} - \frac{\partial \Lambda}{\partial x_\mu} = 0.$$

2.5.1 Relativistic mechanics of a particle

As measured in some inertial reference frame, the instantaneous particle velocity is $\mathbf{v} = (dx/dt, dy/dt, dz/dt)$. It is customary to define noninvariant parameters $\beta = v/c$ and $\gamma = (1 - \beta^2)^{-\frac{1}{2}}$. The definition of the invariant interval ds^2 implies that $d\tau^2 = dt^2(1 - \beta^2)$, where dt is the measured time interval. Hence $\frac{dt}{d\tau} = \gamma$, as defined above. The spatial components of the 4-velocity are $\mathbf{u} = \gamma \mathbf{v}$ and the time component is $u_4 = i\gamma c$. Hence $u_\mu u_\mu = (v^2 - c^2)\gamma^2 = -c^2$, a space-time invariant.

Unlike classical mechanics, the 4-acceleration is not independent of the 4-velocity. Because $u_\mu u_\mu$ is invariant, its τ-derivative must vanish, so that

$$u_\mu \frac{du_\mu}{d\tau} = 0.$$

This implies a relationship between classical velocity and acceleration,

$$\mathbf{v} \cdot \frac{d}{dt}(\gamma \mathbf{v}) - c^2 \frac{d\gamma}{dt} = 0,$$

which is an immediate consequence of the definition of γ.

The variational formalism makes it possible to postulate a relativistic Lagrangian that is Lorentz invariant and reduces to Newtonian mechanics in the classical limit. Introducing a parameter m, the *proper* mass of a particle, or mass as measured in its own instantaneous rest frame, the Lagrangian for a free particle can be postulated to have the invariant form $\Lambda = \frac{1}{2} m u_\mu u_\mu = -\frac{1}{2} mc^2$. The canonical momentum is $p_\mu = m u_\mu$ and the Lagrangian equation of motion is

$$\frac{d}{d\tau} p_\mu = \frac{d}{d\tau}(m u_\mu) = 0,$$

which clearly reduces to Newton's equation when $\beta \to 0$.

As measured in an inertial reference frame, the spatial components of the canonical momentum are $\mathbf{p} = m\gamma \mathbf{v}$ and the time component is $p_4 = im\gamma c$. This can be related to classical quantities by defining the relative energy as $E = m\gamma c^2$ so that $p_4 = \frac{i}{c} E$. Thus p_μ is an energy–momentum vector, for which $p_\mu p_\mu = -m^2 c^2$, a constant for unaccelerated motion. In terms of classical quantities, this invariant norm is expressed by $p^2 - \frac{E^2}{c^2} = -m^2 c^2$, or $E^2 = m^2 c^4 + p^2 c^2$, which implies the famous Einstein formula $E = mc^2$ in the instantaneous rest frame of a particle. In the limit $\beta \to 0$, $E = mc^2 + \frac{1}{2} m v^2 + \cdots$, verifying the classical formula for kinetic energy.

The free-particle Lagrangian Λ is a space-time constant $-\frac{1}{2} mc^2$. If terms are added that are invariant functions of x_μ, the equations of motion become

$$\frac{d}{d\tau} p_\mu = X_\mu$$

defining a covariant Minkowski 4-force $X_\mu = \frac{\partial \Lambda}{\partial x_\mu}$. The canonical momenta are modified if these added terms depend explicitly on the 4-velocity u_μ. It can no longer be assumed that the rest mass is constant.

If m is not constant, the equations of motion are

$$\frac{d}{d\tau}\left(m\frac{dx_\mu}{d\tau}\right) = \frac{dm}{d\tau}\frac{dx_\mu}{d\tau} + m\frac{d^2 x_\mu}{d\tau^2} = X_\mu.$$

On multiplying by $u_\mu = dx_\mu/d\tau$ and summing, and using $u_\mu u_\mu = -c^2$, which implies $u_\mu \frac{d}{d\tau} u_\mu = 0$, this reduces to

$$\frac{d}{d\tau}(mc^2) = -X_\mu u_\mu.$$

The rest mass remains constant if the 4-force and 4-velocity are orthogonal in the Minkowski sense that $X_\mu u_\mu = 0$.

In relativistic theory, energy is a component of the 4-momentum, and is conserved only under particular circumstances. Given $\frac{d}{d\tau} = \gamma \frac{d}{dt}$, the equation of motion for $p_4 = iE/c$ is

$$\gamma \frac{d}{dt}(iE/c) = X_4.$$

Consider a 4-force X_μ that does not change the rest mass m. As shown above, $X_\mu u_\mu = i\gamma c X_4 + \gamma \mathbf{X} \cdot \mathbf{v} = 0$, since $u_4 = i\gamma c$ and $\mathbf{u} = \gamma \mathbf{v}$. Then $X_4 = \frac{i}{c}\mathbf{X} \cdot \mathbf{v}$ for such a force, and

$$\gamma \frac{d}{dt} E = \mathbf{X} \cdot \mathbf{v}.$$

2.5.2 Relativistic motion in an electromagnetic field

The classical electromagnetic force acting on a particle of charge q is the Lorentz force (in Gaussian units)

$$\mathbf{Q} = q\left\{\mathcal{E} + \frac{1}{c}(\mathbf{v} \times \mathcal{B})\right\},$$

where, in terms of scalar and vector potentials ϕ, $\mathbf{A}$,

$$\mathcal{B} = \nabla \times \mathbf{A}, \qquad \mathcal{E} = -\nabla\phi - \frac{1}{c}\frac{\partial \mathbf{A}}{\partial t}.$$

After expanding the triple vector product $\mathbf{v} \times \nabla \times \mathbf{A} = \mathbf{v} \times \mathcal{B}$,

$$\mathbf{Q} = \frac{q}{c}\left\{\frac{\partial}{\partial \mathbf{x}}(\mathbf{A} \cdot \mathbf{v} - \phi c) - \frac{d}{dt}\mathbf{A}\right\}. \tag{2.6}$$

This is of the form

$$\mathbf{Q} = \frac{\partial W}{\partial \mathbf{x}} - \frac{d}{dt}\frac{\partial W}{\partial \mathbf{v}},$$

where $W = \frac{q}{c}(\mathbf{A} \cdot \mathbf{v} - \phi c)$.

Given the electromagnetic 4-vector field $A_\mu = (\mathbf{A}, i\phi)$ and the 4-velocity $u_\mu = (\gamma \mathbf{v}, i\gamma c)$, W is the classical limit of a relativistic invariant $\gamma W = \frac{q}{c} A_\mu u_\mu$. This term augments the free-particle relativistic Lagrangian to give

$$\Lambda = \frac{1}{2} m u_\mu u_\mu + \frac{q}{c} A_\mu u_\mu. \tag{2.7}$$

The canonical momentum is $p_\mu = m u_\mu + \frac{q}{c} A_\mu$. If defined such that $p_4 = \frac{i}{c} E$, the energy is $E = mc^2\gamma + q\phi$, adding electrostatic energy to rest energy mc^2 and relativistic kinetic energy $T = mc^2(\gamma - 1)$. Because $m u_\mu m u_\mu = (p - \frac{q}{c}A)_\mu (p - \frac{q}{c}A)_\mu = -m^2c^2$, energy and momentum are related by $(\mathbf{p} - \frac{q}{c}\mathbf{A})^2 - (mc^2 + T)^2/c^2 = -m^2c^2$, or

$$(mc^2 + T)^2 = \left(\mathbf{p} - \frac{q}{c}\mathbf{A}\right)^2 c^2 + m^2c^4.$$

Written as a formula for iteration, this is

$$T = \left(\mathbf{p} - \frac{q}{c}\mathbf{A}\right)^2 \Big/ (2m + T/c^2).$$

The Lagrangian equations of motion are

$$\frac{d}{d\tau}\left(m u_\mu + \frac{q}{c} A_\mu\right) = \frac{q}{c}\frac{\partial}{\partial x_\mu}(A_\nu u_\nu).$$

Using $\frac{d}{d\tau} = \gamma \frac{d}{dt}$, these equations can be written as

$$\gamma \frac{d}{dt}(m u_\mu) = X_\mu = \frac{q}{c}\left\{\frac{\partial}{\partial x_\mu}(A_\nu u_\nu) - \gamma \frac{d}{dt} A_\mu\right\}. \tag{2.8}$$

Comparing the classical force $\mathbf{Q}$, given above, the spatial components of X_μ are $\mathbf{X} = \gamma \mathbf{Q}$, verifying the historical fact that Maxwell's theory is covariant under Lorentz transformations.

3
Applied mathematics

The principal references for this chapter are:

[77] Dennis, J.E. and Schnabel, R.B. (1983). *Numerical Methods for Unconstrained Optimization and Nonlinear Equations* (Prentice-Hall, Englewood Cliffs, New Jersey).

[125] Fletcher, R. (1987). *Practical Methods of Optimization*, 2nd edition (Wiley, New York).

[168] Hestenes, M.R. (1966). *Calculus of Variations and Optimal Control Theory* (Wiley, New York).

[332] Pulay, P. (1987). Analytical derivative methods in quantum chemistry, *Adv. Chem. Phys.* **69**, 241–286.

[439] Zerner, M.C. (1989). Analytic derivative methods and geometry optimization, in *Modern Quantum Chemistry*, eds. A. Szabo and N.S. Ostlund (McGraw-Hill, New York), pp. 437–458.

Before undertaking the major subject of variational principles in quantum mechanics, the present chapter is intended as a brief introduction to the extension of variational theory to linear dynamical systems and to classical optimization methods. References given above and in the Bibliography will be of interest to the reader who wishes to pursue this subject in fields outside the context of contemporary theoretical physics and chemistry. The specialized subject of optimization of molecular geometries in theoretical chemistry is treated here in some detail.

3.1 Linear systems

Any multicomponent system whose dynamical behavior is governed by coupled linear equations can be modelled by an effective Lagrangian, quadratic in the system variables. Hamilton's variational principle is postulated to determine the time behavior of the system. A dynamical model of some system of interest is valid if it satisfies the same system of coupled equations. This makes it possible, for example, to construct an electrical circuit model of a mechanical system, or to reduce either to a computer model with the appropriate choice of parameters.

For example [146], a system of interconnected electrical circuits and a mechanical system of masses connected by springs satisfy the same linear equations if system parameters are related by the following definitions:

Symbol	Electrical circuit	Masses on springs
I	current	displacement
M	inductance	reciprocal mass tensor
R	resistance	viscous force
C	capacitance	inverse spring constant
E	external emf	driving force

The Lagrangian for this linear system is

$$L = \frac{1}{2}\sum_{j,k} M_{jk}\dot{I}_j\dot{I}_k - \frac{1}{2}\sum_j \frac{1}{C_j}I_j^2 + \sum_j \dot{E}_j(t)I_j.$$

A dissipation function adds a nonintegrable term

$$\delta W = -\sum_j R_j\dot{I}_j\delta I_j$$

to δL. The implied equations of motion are

$$\sum_k M_{jk}\ddot{I}_k + R_j\dot{I}_j + I_j/C_j = \dot{E}_j(t).$$

3.2 Simplex interpolation

In many practical applications of nonlinear optimization, a linear approximation is iterated until some vector of system parameters has negligible norm. At each stage of such a process, previous steps provide a set of $m \leq n$ trial vectors in an n-dimensional parameter space, each associated with an output vector of gradients of the quantity to be optimized. The next iterative step is facilitated if a linear combination of such vectors can be found that produces an output vector of minimum norm.

Linear interpolation can be described geometrically in terms of an *m-simplex* in the n-dimensional parameter space. An m-simplex in a hyperspace of dimension $n \geq m$ is a set of $m+1$ points that do not lie in a subspace of dimension less than m. For example, a triangle is a 2-simplex and a tetrahedron is a 3-simplex. The interior of a simplex is the set of points $\mathbf{X} = \sum_i \mathbf{x}_i\lambda_i$, such that $0 \leq \lambda_i \leq 1$ and $\sum_i \lambda_i = 1$.

A typical optimization problem is to interpolate a vectorial function $\mathbf{f}(\mathbf{x})$ to a given vector $\mathbf{F}$ by a least-square fit $\mathbf{F} \simeq \sum_i \mathbf{f}_i \lambda_i$, where $\mathbf{f}_i = \mathbf{F}(\mathbf{x}_i)$. The variational problem is to minimize $\frac{1}{2}|\sum_i \mathbf{f}_i \lambda_i - \mathbf{F}|^2$, subject to the linear constraint $\sum_i \lambda_i = 1$, using a Lagrange multiplier μ. The variational functional is $I_\mu = \frac{1}{2}\sum_{i,j}(\mathbf{f}_i \lambda_i - \mathbf{F}) \cdot (\mathbf{f}_j \lambda_j - \mathbf{F}) - (\sum_i \lambda_i - 1)\mu$. The variational equations, for $i = 1, \ldots, m$, are

$$\sum_j \mathbf{f}_i \cdot \mathbf{f}_j \lambda_j - \mu = \mathbf{f}_i \cdot \mathbf{F}; \qquad \sum_j \lambda_j = 1.$$

The simplest possible example is linear interpolation in a 1-simplex (x_0, x_1). The system of linear equations is

$$\begin{aligned} f_0 f_0 \lambda_0 + f_0 f_1 \lambda_1 - \mu &= f_0 F \\ f_1 f_0 \lambda_0 + f_1 f_1 \lambda_1 - \mu &= f_1 F \\ \lambda_0 + \lambda_1 &= 1. \end{aligned}$$

The solution $\lambda_0 = \frac{f_1 - F}{f_1 - f_0}$; $\lambda_1 = \frac{F - f_0}{f_1 - f_0}$; $\mu = 0$ determines the coefficients in the interpolation formula

$$X = \frac{x_0(f_1 - F) + x_1(F - f_0)}{f_1 - f_0}.$$

3.2.1 Extremum in n dimensions

For an extremum in n dimensions, $\delta W(\mathbf{q}) = 0$, where $\mathbf{q}$ is a vector of n generalized coordinates. Generalized gradients, defined by $p_i = \frac{\partial W}{\partial q_i}$ for $i = 1, \ldots, n$, must vanish at an extremum. A linear interpolating function $\mathbf{P} = \sum_i \mathbf{p}_i \lambda_i$ is to be interpolated to $\mathbf{P} \simeq \mathbf{0}$, at the corresponding point $\mathbf{Q} = \sum_i \mathbf{q}_i \lambda_i$ in the coordinate hyperspace. The variational equations for the simplex method are

$$\sum_j \mathbf{p}_i \cdot \mathbf{p}_j \lambda_j - \mu = 0; \qquad \sum_j \lambda_j = 1.$$

On solving these equations for the coefficients $\{\lambda_j\}$, the solution of minimum norm is the interpolated gradient vector $\mathbf{P}$, such that $\mu = |\mathbf{P}|^2 \simeq 0$, at the interpolated coordinate vector $\mathbf{Q}$. The Lagrange multiplier μ in this method provides an estimate of the residual error.

At each step of an iterative method for nonlinear optimization, the subsequent coordinate step $\Delta\mathbf{q}$ must be estimated. The vector $\mathbf{Q}$, interpolated in a selected m-simplex of prior coordinate vectors, must be combined with an iterative estimate of the component of $\Delta\mathbf{q}$ orthogonal to the hyperplane of the simplex.

A Hessian matrix

$$F_{ij} = \frac{\partial^2 W}{\partial \mathbf{q}_i \partial \mathbf{q}_j},$$

is defined by small displacements about a given point, such that $\Delta\mathbf{p} = F\Delta\mathbf{q}$. A generalization of Newton's method can be used to estimate the location of a zero value of a vector from its given value and derivative. If an approximation to F^{-1} is maintained and updated at each iterative step, then

$$\Delta\mathbf{q} = \Delta\mathbf{q}_\perp + \sum_i \mathbf{q}_i \lambda_i,$$

where

$$\Delta\mathbf{q}_\perp = -F^{-1}\mathbf{p} + \sum_{i,j\leq m} (F^{-1}\mathbf{p} \cdot \mathbf{q}_i)[(\mathbf{q} \cdot \mathbf{q})]_{ij}^{-1} \mathbf{q}_j.$$

Alternatively, the interpolation and extrapolation steps can be combined, using the formula [75]

$$\mathbf{q} = \mathbf{Q} - F^{-1}\mathbf{P} = \sum_i (\mathbf{q}_i - F^{-1}\mathbf{p}_i)\lambda_i.$$

3.3 Iterative update of the Hessian matrix

Quadratic expansion of a function $W(\mathbf{q})$ about a local extremum takes the form

$$W \simeq W^0 + \frac{1}{2} \sum_{i,j} F^0_{i,j} \Delta q_i \Delta q_j.$$

At a general point, displaced from the extremum, gradients $p_i = \frac{\partial W}{\partial q_i}$ do not vanish. Newton's formula estimates a displacement toward the extremum $\Delta\mathbf{q} = -\mathbf{G}\mathbf{p}^0$, where $\mathbf{G}$ is the inverse of the Hessian matrix $\mathbf{F}$. Since $\mathbf{F}$ is not constant in a nonlinear problem, if it cannot be computed directly it must be deduced from an initial estimate followed by iterative updates. In many circumstances, the gradients at a general point can be computed directly or estimated with useful accuracy. Then each successive coordinate increment $\Delta\mathbf{q}$ is associated with a gradient increment $\Delta\mathbf{p}$. If the Hessian matrix is known, $\mathbf{F}\Delta\mathbf{q} = \Delta\mathbf{p}$ for sufficiently small increments. Given an estimated F^0, it must in general be updated to be consistent with the computed gradient increment $\Delta\mathbf{p}$. This implies a linear formula for the increment of F, $\Delta F \Delta\mathbf{q} = \Delta\mathbf{p} - F^0\Delta\mathbf{q}$. Since this provides only n equations for the n^2 elements of ΔF, the incremental matrix cannot be determined from data obtained in a single iterative step.

The practical problem is to find a way to use the information obtained in each iterative step without making unjustified changes of the Hessian. This could be accomplished if the Hessian were projected onto two orthogonal subspaces, one of which corresponds to a subset of m linearly independent increments $\Delta\mathbf{q}$ and the corresponding $\Delta\mathbf{p}$. A *rank-m* (*Rm*) update in the vector space spanned by these coordinate increments is defined such that $\Delta\mathbf{q}^\dagger\Delta F\Delta\mathbf{q} = \Delta\mathbf{q}^\dagger(\Delta\mathbf{p} - F^0\Delta\mathbf{q})$. This condition is satisfied by

$$\Delta F(Rm) = \Delta\mathbf{p}(\Delta\mathbf{q}^\dagger\Delta\mathbf{p})^{-1}\Delta\mathbf{q}^\dagger(\Delta\mathbf{p} - F^0\Delta\mathbf{q})(\Delta\mathbf{p}^\dagger\Delta\mathbf{q})^{-1}\Delta\mathbf{p}^\dagger.$$

Alternatively, the inverse matrix G can be updated directly,

$$\Delta G(Rm) = \Delta\mathbf{q}(\Delta\mathbf{p}^\dagger\Delta\mathbf{q})^{-1}\Delta\mathbf{p}^\dagger(\Delta\mathbf{q} - G^0\Delta\mathbf{p})(\Delta\mathbf{q}^\dagger\Delta\mathbf{p})^{-1}\Delta\mathbf{q}^\dagger.$$

These formulas update the rank-m projection of F^0 or G^0, using the nonhermitian projection operator $\mathcal{P}_m = \Delta\mathbf{p}(\Delta\mathbf{q}^\dagger\Delta\mathbf{p})^{-1}\Delta\mathbf{q}^\dagger$, such that $\mathcal{P}_m^\dagger\Delta\mathbf{q} = \Delta\mathbf{q}$, $\mathcal{P}_m\Delta\mathbf{p} = \Delta\mathbf{p}$, and $\mathcal{P}_m\mathcal{P}_m = \mathcal{P}_m$. This operator projects onto the m-dimensional vector space spanned by the specified set of gradient vectors. The *Rm* update has the undesirable property of altering the complementary projection of the updated matrix.

3.3.1 The BFGS algorithm

The BFGS (Broyden [42], Fletcher [124], Goldfarb [145], Shanno [379]) algorithm is an update procedure for the Hessian matrix that is widely used in iterative optimization [125]. The simpler *Rm* update takes the form

$$F = \Delta F_{mm} + F^0 - \mathcal{P}_m F^0 \mathcal{P}_m^\dagger,$$

where $\Delta F_{mm}\Delta\mathbf{q} = \Delta\mathbf{p}$. Replacing this by a form that leaves the complementary projection of F^0 unchanged,

$$F = \Delta F_{mm} + (I - \mathcal{P}_m)F^0(I - \mathcal{P}_m^\dagger),$$

the BFGS update of F^0 is

$$\Delta F(BFGS) = \Delta\mathbf{p}(\Delta\mathbf{q}^\dagger\Delta\mathbf{p})^{-1}\Delta\mathbf{q}^\dagger(\Delta\mathbf{p} + F^0\Delta\mathbf{q})(\Delta\mathbf{p}^\dagger\Delta\mathbf{q})^{-1}\Delta\mathbf{p}^\dagger - \Delta\mathbf{p}(\Delta\mathbf{q}^\dagger\Delta\mathbf{p})^{-1}\Delta\mathbf{q}^\dagger F^0 - F^0\Delta\mathbf{q}(\Delta\mathbf{p}^\dagger\Delta\mathbf{q})^{-1}\Delta\mathbf{p}^\dagger.$$

Alternatively, the BFGS update of G^0 is

$$\Delta G(BFGS) = \Delta\mathbf{q}(\Delta\mathbf{p}^\dagger\Delta\mathbf{q})^{-1}\Delta\mathbf{p}^\dagger(\Delta\mathbf{q} + G^0\Delta\mathbf{p})(\Delta\mathbf{q}^\dagger\Delta\mathbf{p})^{-1}\Delta\mathbf{q}^\dagger - \Delta\mathbf{q}(\Delta\mathbf{p}^\dagger\Delta\mathbf{q})^{-1}\Delta\mathbf{p}^\dagger G^0 - G^0\Delta\mathbf{p}(\Delta\mathbf{q}^\dagger\Delta\mathbf{p})^{-1}\Delta\mathbf{q}^\dagger.$$

For exact infinitesimal increments, the matrix

$$dp^\dagger dq = dq^\dagger dp = dq^\dagger F\, dq$$

is symmetric and nonsingular at an extremum of W. When $m > 1$, the matrices $\Delta\mathbf{p}^\dagger\Delta\mathbf{q}$ and $\Delta\mathbf{q}^\dagger\Delta\mathbf{p}$ are not necessarily symmetric or even nonsingular when evaluated with approximate gradients away from an extremum. Any practical algorithm must modify these matrices to remove singularities.

3.4 Geometry optimization for molecules

Quantum mechanical calculations of the electronic structure of molecules for fixed nuclear coordinates involve lengthy calculations even using the sophisticated computational methods that have evolved over half a century of computational quantum chemistry. A principal output of such calculations is the variational energy as a function of the nuclear coordinates. Current methodology makes it possible to compute the energy gradients or effective forces as well as the energy itself. For chemical applications, equilibrium geometries and the coordinates of transition states must be deduced from such data [357, 332, 439].

For such applications of classical optimization theory, the data on energy and gradients are so computationally expensive that only the most efficient optimization methods can be considered, no matter how elaborate. The number of quantum chemical wave function calculations must absolutely be minimized for overall efficiency. The computational cost of an update algorithm is always negligible in this context. Data from successive iterative steps should be saved, then used to reduce the total number of steps. Any algorithm dependent on line searches in the parameter hyperspace should be avoided.

Molecular geometry can be specified either in Cartesian coordinates $\mathbf{x}_a$ for each of N nuclei, or in generalized internal coordinates q_k, where $k = 1, \ldots, 3N - 6$ for a general polyatomic molecule. Neglecting kinetic energy of nuclear motion because of the large nuclear masses, six generalized coordinates corresponding to translation and rotation are subject to no internal forces, and can be fixed or removed from the internal coordinates considered in geometry optimization. Cartesian coordinates are subject to six constraint conditions, which may be imposed directly or indirectly using Lagrange multipliers. This number reduces to five for a linear molecule, since rotation about the molecular axis is not defined.

An equilibrium state is defined for generalized coordinates such that the total energy $E(\{q\})$ is minimized. The energy gradients $p = \frac{\partial E}{\partial q}$ are forces with reversed sign. Coordinate and gradient displacements from m successive iterations are saved as $m \times n$ column matrices, $\Delta\mathbf{q}_k$ and $\Delta\mathbf{p}_k$, respectively. The Hessian matrix is $F_{ij} = \frac{\partial^2 E}{\partial q_i \partial q_j}$, such that Newton's extrapolation formula is $\Delta\mathbf{q} = -\mathbf{G}\mathbf{p}^0$, where $\mathbf{G} = \mathbf{F}^{-1}$.

In *quasi-Newton* methods, a parametrized estimate of F or G is used initially, then updated at each iterative step. Saved values of q, p can be used in standard algorithms such as BFGS, described above.

3.4.1 The GDIIS algorithm

This method [75] uses simplex interpolation combined with an update step such as BFGS. The variational functional is

$$I_\mu = \frac{1}{2}\left|\sum_{k=0}^{m} \mathbf{p}^k c_k\right|^2 - \left(\sum_{k=0}^{m} c_k - 1\right)\mu.$$

The variational equations determined by $\delta I_\mu = 0$ are

$$\sum_j \mathbf{p}^k \cdot \mathbf{p}^j c_j - \mu = 0; \qquad \sum_j c_j = 1.$$

The interpolated gradient vector is $\bar{\mathbf{p}} = \sum_k \mathbf{p}^k c_k$ and the interpolated coordinate vector is $\bar{\mathbf{q}} = \sum_k \mathbf{q}^k c_k$. In the GDIIS algorithm, a currently updated estimate of the inverse Hessian is used to estimate a coordinate step based on these interpolated vectors. This gives

$$\Delta\bar{\mathbf{q}} = \mathbf{q}^{upd} - \bar{\mathbf{q}} = -(G^0 + \Delta G)\bar{\mathbf{p}}$$

or, equivalently, with $G = G^0 + \Delta G$,

$$\mathbf{q}^{upd} = \bar{\mathbf{q}} - G\bar{\mathbf{p}} = \sum_k (\mathbf{q}^k - G\mathbf{p}^k) c_k.$$

In test calculations [75] this algorithm was found to produce rapid convergence. When combined with single-step ($m = 1$) BFGS update of the inverse Hessian, this is a very efficient algorithm.

The GDIIS algorithm can locate saddle points (transition states), because it searches specifically for a point at which all gradients vanish, independently of the sign of the second derivative. A reaction path can be followed by selecting the eigenvector of G that belongs to the eigenvalue of greatest magnitude [332].

3.4.2 The BERNY algorithm

This algorithm, standard in the widely used GAUSSIAN program system, is a rank-m update of the Hessian matrix, in an orthonormal basis [356]. A basis of unit vectors is constructed in the m-dimensional vector space spanned by the increments $\Delta\mathbf{q}$. For $k = 1, m$, define

$$\mathbf{d}^k = \Delta\mathbf{q}^k - \sum_{j=1}^{k-1} \mathbf{e}^j (\mathbf{e}^j \cdot \Delta\mathbf{q}^k).$$

Then orthonormal unit vectors are constructed such that

$$\mathbf{e}^k = \mathbf{d}^k (\mathbf{d}^k \cdot \mathbf{d}^k)^{-\frac{1}{2}}.$$

A significant advantage of this procedure is that nearly linearly dependent vectors can be eliminated at this stage, simply reducing the value of m. The resulting unit vectors define an $n \times m$ column matrix. Using these unit vectors, the algorithm solves an $m \times m$ system of linear equations in the **e**-space,

$$(\Delta \mathbf{q}^\dagger \mathbf{e})(\mathbf{e}^\dagger \bar{F} \mathbf{e}) = (\Delta \mathbf{p}^\dagger \mathbf{e}).$$

The matrix $(\Delta \mathbf{q}^\dagger \mathbf{e})$ is lower-triangular by construction. For the index range $j \le i \le m$, solution matrix elements are determined by

$$(\mathbf{e}^{i\dagger} \bar{F} \mathbf{e}^j) = (\Delta \mathbf{q}^{i\dagger} \mathbf{e}^i)^{-1} \left[(\Delta \mathbf{p}^{i\dagger} \mathbf{e}^j) - \sum_{k=1}^{i-1} (\Delta \mathbf{q}^{i\dagger} \mathbf{e}^k)(\mathbf{e}^{k\dagger} \bar{F} \mathbf{e}^j) \right].$$

The Hessian matrix $F^0 + \Delta F$ is updated and symmetrized, using

$$\Delta F = \sum_{i,j} \mathbf{e}^i [(\mathbf{e}^{i\dagger} \bar{F} \mathbf{e}^j)_{j \le i} + (\mathbf{e}^{j\dagger} \bar{F} \mathbf{e}^i)_{j > i} - (\mathbf{e}^{i\dagger} F^0 \mathbf{e}^j)] \mathbf{e}^{j\dagger}.$$

Several aspects of the BERNY algorithm have a somewhat inconsistent mathematical basis. A modified algorithm may be more efficient. The Hessian matrix can be constructed directly from gradient vectors but not from coordinate vectors. The defining relation $\mathbf{F}\Delta\mathbf{q} = \Delta\mathbf{p}$ implies that an Hermitian matrix $\mathbf{F}$ must have the form $\Delta\mathbf{p} A \Delta\mathbf{p}^\dagger$. In principle, the unit vectors of the BERNY algorithm should be in the $\Delta\mathbf{p}$ space. However, the orthonormalization process ensures numerical stability by systematizing rejection of redundant coordinate increments $\Delta\mathbf{q}$. One might then propose an alternative algorithm, in which this basis is used to update the inverse Hessian $\mathbf{G}$. Because $\mathbf{G}\Delta\mathbf{p} = \Delta\mathbf{q}$, this matrix has the formal expansion $\Delta\mathbf{q} B \Delta\mathbf{q}^\dagger$. The equations to be solved are

$$(\Delta \mathbf{p}^\dagger \mathbf{e})(\mathbf{e}^\dagger \bar{G} \mathbf{e}) = (\Delta \mathbf{q}^\dagger \mathbf{e}).$$

In these equations, the matrix $(\Delta\mathbf{p}^\dagger\mathbf{e})$ is no longer triangular. However, the additional computational effort may be unimportant if the number of iterative steps can be reduced, thus saving energy and gradient evaluations. The matrix $\mathbf{e}^\dagger \bar{G} \mathbf{e}$ is to be symmetrized before updating $\mathbf{G}$. The incremental matrix is

$$\Delta G = \sum_{i,j} \mathbf{e}^i \left[\frac{1}{2}(\mathbf{e}^{i\dagger} \bar{G} \mathbf{e}^j) + \frac{1}{2}(\mathbf{e}^{j\dagger} \bar{G} \mathbf{e}^i) - (\mathbf{e}^{i\dagger} G^0 \mathbf{e}^j) \right] \mathbf{e}^{j\dagger}.$$

II

Bound states in quantum mechanics

This part introduces variational principles relevant to the quantum mechanics of bound stationary states. Chapter 4 covers well-known variational theory that underlies modern computational methodology for electronic states of atoms and molecules. Extension to condensed matter is deferred until Part III, since continuum theory is part of the formal basis of the multiple scattering theory that has been developed for applications in this subfield. Chapter 5 develops the variational theory that underlies independent-electron models, now widely used to transcend the practical limitations of direct variational methods for large systems. This is extended in Chapter 6 to time-dependent variational theory in the context of independent-electron models, including linear-response theory and its relationship to excitation energies.

4

Time-independent quantum mechanics

The principal references for this chapter are:

[74] Courant, R. and Hilbert, D. (1953). *Methods of Mathematical Physics* (Interscience, New York).

[101] Epstein, S.T. (1974). *The Variation Method in Quantum Chemistry* (Academic Press, New York).

[180] Hurley, A.C. (1976). *Electron Correlation in Small Molecules* (Academic Press, New York).

[242] Merzbacher, E. (1961). *Quantum Mechanics* (Wiley, New York).

[365] Schrödinger, E. (1926). Quantisiering als Eigenwertproblem, *Ann. der Physik* **81**, 109–139.

[394] Szabo, A. and Ostlund, N. S. (1982). *Modern Quantum Chemistry; Introduction to Advanced Electronic Structure Theory* (McGraw-Hill, New York).

In 1926, Schrödinger [365] recognized that the variational theory of elliptical differential equations with fixed boundary conditions could produce a discrete eigenvalue spectrum in agreement with the energy levels of Bohr's model of the hydrogen atom. This conceptually startling amalgam of classical ideas of particle and field turned out to be correct. Within a few years, the new wave mechanics almost completely replaced the *ad hoc* quantization of classical mechanics that characterized the "old" quantum theory initiated by Bohr. Although the matrix mechanics of Heisenberg was soon shown to be logically equivalent, the variational wave theory became the standard basis of methodology in the physics of electrons.

The nonrelativistic Schrödinger theory is readily extended to systems of N interacting electrons. The variational theory of finite N-electron systems (atoms and molecules) is presented here. In this context, several important theorems that follow from the variational formalism are also derived.

Hartree atomic units will be used here. In these units, the unit of action $\hbar$, the mass m of the electron, and the magnitude e of the electronic charge $-e$ are all set equal to unity. The velocity of light c is $1/\alpha$, where α is the fine-structure constant.

The unit of length is a_0, the first Bohr radius of atomic hydrogen. The Hartree unit of energy is e^2/a_0, approximately 27.212 electron volts.

4.1 Variational theory of the Schrödinger equation

The Schrödinger equation for one electron is

$$\{\hat{t} + v(\mathbf{r}) - \epsilon\}\psi(\mathbf{r}) = 0,$$

where $\hat{t} = -\frac{1}{2}\nabla^2$, the kinetic energy operator of Schrödinger. For historical reasons, this is written as

$$\{\mathcal{H} - \epsilon\}\psi = 0,$$

where $\mathcal{H} = \hat{t} + v(\mathbf{r})$ is the Hamiltonian operator of the theory. It is assumed that physically meaningful potential functions $v(\mathbf{r})$ vanish for large $r \to \infty$. Solutions ψ are required to be bounded in $\mathcal{R}^3$, 3-dimensional Euclidean space, and to be continuous with continuous gradients except at Coulomb singularities of the potential function. These conditions define the Hilbert space of trial functions for the variational theory. Eigenfunctions of the one-electron Hamiltonian that lie in this Hilbert space will be called *orbital* wave functions here.

Because $\mathcal{H}$ is Hermitian, the energy eigenvalues ϵ are real numbers. For any physically realizable potential function $v(\mathbf{r})$, there is a lowest eigenvalue ϵ_0. For bound states, discrete eigenvalues $\epsilon < 0$ are determined by the condition that the wave function ψ must vanish as $r \to \infty$. Continuum states, with $\epsilon \geq 0$, are bounded, oscillatory functions at large r, but must be regular at the coordinate origin. The orbital Hilbert space is characterized by a scalar product $(i|j) = \int d^3\mathbf{r}\psi_i^*(\mathbf{r})\psi_j(\mathbf{r})$.

4.1.1 Sturm–Liouville theory

It is generally true that the normalized eigenfunctions of an Hermitian operator such as the Schrödinger $\mathcal{H}$ constitute a complete orthonormal set in the relevant Hilbert space. A completeness theorem is required in principle for each particular choice of $v(\mathbf{r})$ and of boundary conditions. To exemplify such a proof, it is helpful to review classical Sturm–Liouville theory [74] as applied to a homogeneous differential equation of the form

$$L[f(x)] + \lambda\rho(x)f(x) = 0,$$

where $L[f] = (pf')' - qf$, for continuous p, p', q with $\rho, p > 0$ in the interval $x_0 \leq x \leq x_1$. The Hilbert space of solutions $f(x)$ is defined by real-valued continuous functions with continuous first derivatives over the interval $x_0 \leq x \leq x_1$, subject to homogeneous boundary conditions (specified logarithmic derivatives) at the end-points x_0, x_1.

A version of Green's theorem follows from partial integration of the symmetric integral

$$\int (pf_1' f_2' + qf_1 f_2)\, dx = -\int f_2 L[f_1]\, dx + pf_1' f_2|_{x_0}^{x_1},$$

which implies

$$\int (f_1(x)L[f_2(x)] - f_2(x)L[f_1(x)])\, dx = p(f_1 f_2' - f_1' f_2)|_{x_0}^{x_1}.$$

For homogeneous boundary conditions, the logarithmic derivatives f_1'/f_1 and f_2'/f_2 are equal at both end-points x_0, x_1. Hence the integrated term vanishes, and the differential expression $L[f]$ is *self-adjoint* with these boundary conditions. The weighting function ρ can be eliminated by converting to $u = \rho^{\frac{1}{2}} f$. Then $\Lambda[u] = (Pu')' - Qu$, where

$$P = \frac{p}{\rho}; \qquad Q = \frac{q}{\rho} - \rho^{-\frac{1}{2}} \frac{d}{dx}\left(p \frac{d}{dx} \rho^{-\frac{1}{2}}\right),$$

and

$$\Lambda[u(x)] + \lambda u(x) = 0.$$

A Green function is defined as a solution of the inhomogeneous equation (using a Dirac delta-function)

$$\Lambda[G(x, \xi)] = -\delta(x, \xi),$$

subject to the specified boundary conditions. Then

$$u(x) = \lambda \int_{x_0}^{x_1} G(x, \xi) u(\xi)\, d\xi$$

is a solution of the differential equation with these boundary conditions. Because $\Lambda[u]$ is self-adjoint, u satisfies an homogeneous integral equation with a symmetrical kernel [74].

Starting from an assumed minimum eigenvalue λ_0, for state $u_0(x)$, a nondecreasing sequence can be built up one by one. A global constant can be added so that $\lambda_0 > 0$, and then removed when all eigenvalues are determined. This construction uses the theorem that the maximum value of $(u|G|u) = \int\int u(s)G(s,t)u(t)\, dt\, ds$, subject to $(u|u) = 1$, is given by the eigenfunction u_0 corresponding to the minimum eigenvalue λ_0. This follows from $\delta[(u|G|u) - \mu((u|u) - 1)] = 0$, for which the Euler equation is the homogeneous integral equation

$$\int G(s,t)u(t)\,dt = \mu u(s),$$

where the Lagrange multiplier $\mu = \lambda^{-1}$. The maximized value $(u|G|u)$ is $\mu(u|u) = \lambda_0^{-1}$, selecting the lowest eigenvalue. Given u_0 and λ_0, consider the modified Green function $G_1(s,t) = G(s,t) - u_0(s)\lambda_0^{-1}u_0(t)$. If $(u|G_1|u)$ is maximized, subject to the orthogonality constraint $(u|u_0) = 0$ and to normalization $(u|u) = 0$, the same procedure obtains u_1 and the eigenvalue λ_1. Equivalently, u_{n+1} maximizes $(u|G|u)$ subject to orthogonality to all eigenfunctions u_i with $i \leq n$. Thus an entire countable sequence of eigenvalues and orthonormal eigenfunctions can be constructed.

If the full set of eigenfunctions is complete, any function χ in the Hilbert space that is orthogonal to all eigenfunctions must vanish identically. To prove this, suppose that some function χ exists such that $(u_i|\chi) = 0, i \leq n$, but $(\chi|\chi) = 1$ and $\lambda(\chi|G|\chi) = 1$ for finite positive λ. The construction given above develops the expansion

$$G(s,t) = \lim_{n\to\infty} \sum_{i=0}^{n} u_i(s)u_i(t)/\lambda_i,$$

which can be proven to converge uniformly for a positive-definite sequence of eigenvalues (Mercer's theorem) [74]. Because the summation in $(\chi|G|\chi) = \lim_{n\to\infty}\sum_{i=0}^{n}(\chi|u_i)(u_i|\chi)/\lambda_i$ vanishes term by term, due to the orthogonality condition, the sum must converge to zero. Since λ is assumed to be finite, and $(\chi|\chi) = \lambda(\chi|G|\chi)$, this contradicts the hypothesis that $\chi(x)$ is normalizable to unity, implying that $\chi(x)$ must vanish identically for $x_0 \leq x \leq x_1$ if it satisfies the defining properties for functions $u(x)$ in the Hilbert space. This proves completeness of the orthonormal set of eigenfunctions in the designated Hilbert space [74].

4.1.2 Idiosyncracies of the Schrödinger equation

In time-dependent quantum mechanics, wave functions contain an inherently complex time-factor. There is a significant notational advantage in treating wave functions as complex fields even in bound-state applications of the theory. In particular, complex spherical harmonics are assumed in the vector-coupling formalism required for systematic treatment of orbital and spin angular momentum. The Bloch waves appropriate to regular periodic solids are also inherently complex fields. The extension from real fields is straightforward, using a notation for matrix elements such that

$$(i|j) = \int d^3\mathbf{r}\,\psi_i^*(\mathbf{r})\psi_j(\mathbf{r}); \qquad (i|\mathcal{H}|j) = \int d^3\mathbf{r}\,\psi_i^*(\mathbf{r})\mathcal{H}\psi_j(\mathbf{r}).$$

This implies $(j|i)^* = (i|j)$. Orthonormality is expressed by $(i|j) = \delta_{ij}$. If $\mathcal{H}$ is Hermitian (self-adjoint), $(i|\mathcal{H}|j) - (j|\mathcal{H}|i)^* = (\epsilon_j - \epsilon_i^*)(i|j) = 0$. Since $(i|i) = 1$, ϵ_i must be real. If $j \neq i$, either $\epsilon_j = \epsilon_i$ or $(i|j) = 0$.

For a single electron moving in a fixed potential function $v(\mathbf{r})$ the coordinate range is infinite in $\mathcal{R}^3$ and positive energy eigenvalues lie in a continuum. This spectrum can be discretized by placing the system in a polyhedral or spherical box. The first alternative is appropriate to solid-state physics. Assuming a regular periodic lattice composed of space-filling polyhedral atomic cells, periodic boundary conditions are appropriate on each cell boundary. This situation will be discussed in detail in the chapter on multiple scattering theory. A spherical boundary is appropriate to an isolated atom. However, the relevant physics is not that of bound states but rather the theory of electron scattering, which will also be discussed in a separate chapter. For the variational theory of bound states, box normalization makes it possible to consider completeness in terms of an artificially discrete eigenvalue spectrum, but is not necessarily the most direct way to achieve practical completeness in variational calculations. A widely used practical alternative is to retain the typical bound-state boundary condition that functions vanish for $r \to \infty$, but to use discrete basis sets constructed from spherical harmonics, powers of r, and exponential or Gaussian exponential factors in radial wave functions. In general, practical completeness can be achieved within some effective convergence radius that depends on the choice of decay parameters for such exponential functions. If the boundary conditions allow a true continuum, then the subset of bound states cannot be complete. The bound-state sum must be augmented by an integral over a parameter associated with the continuum.

The usual Hilbert-space requirement of continuous gradients is not appropriate to Coulombic point-singularities of the potential function $v(\mathbf{r})$ [196]. This is illustrated by the cusp behavior of hydrogenic bound-state wave functions, for which the Hamiltonian operator is

$$\mathcal{H} = \hat{t} + v(\mathbf{r}) = \hat{t} - Z/r.$$

It is convenient to use spherical polar coordinates (r, θ, ϕ) for any spherically symmetric potential function $v(r)$. The surface spherical harmonics Y_ℓ^m satisfy Sturm–Liouville equations in the angular coordinates and are eigenfunctions of the orbital angular momentum operator $\hat{\ell}$ such that

$$\hat{\ell} \cdot \hat{\ell} Y_\ell^m = \ell(\ell + 1) Y_\ell^m$$
$$\hat{\ell}_z Y_\ell^m = m Y_\ell^m.$$

The kinetic energy operator takes the form

$$\hat{t} = -\frac{1}{2r^2}\frac{\partial}{\partial r}\left(r^2\frac{\partial}{\partial r}\right) + \frac{\hat{\ell}^2}{2r^2}.$$

For a spatial wave function of the form

$$\psi(r,\theta,\phi) = r^{-1}u(r)Y_\ell^m(\theta,\phi),$$

the radial wave equation is

$$\left\{-\frac{d^2}{2dr^2} + \frac{\ell(\ell+1)}{2r^2} + v(r) - \epsilon\right\}u(r) = 0.$$

A radial wave function is normalized so that $(u|u) = \int_0^\infty u^*(r)u(r)dr = 1$.

A solution regular at the origin must have the form

$$u(r) = r^{\ell+1}(u_0 + ru_1 + \cdots).$$

The leading terms in the radial equation for $r^{-1}u$ are

$$r^{\ell-2}\left\{\frac{1}{2}[-(\ell+1)\ell + \ell(\ell+1)]u_0\right\} +$$
$$r^{\ell-1}\left\{\frac{1}{2}[-(\ell+2)(\ell+1) + \ell(\ell+1)]u_1 - Zu_0\right\} + \cdots.$$

The first term vanishes due to the leading factor in $u(r)$. To avoid a singularity, the second term must vanish for $\ell = 0$. This implies that

$$u_1 = -\frac{Z}{\ell+1}u_0,$$

which for $\ell = 0$ is the Coulomb cusp condition. An external potential that is regular at the origin cannot affect this $r^{\ell-1}$ term even if $\ell > 0$. Hence both even and odd powers of r must be present in the wave function, which implies a discontinuous gradient for $\ell = 0$. The orbital Hilbert space must be defined so that the $\ell = 0$ angular components of wave functions satisfy this cusp condition when expanded about Coulomb singularities in spherical polar coordinates.

4.1.3 Variational principles for the Schrödinger equation

Schrödinger [365] defined the kinetic energy functional as the positive-definite form $T = \frac{1}{2}\int_\tau d^3\mathbf{r}\ \nabla\psi^*(\mathbf{r})\cdot\nabla\psi(\mathbf{r})$, valid in a finite volume τ enclosed by a surface σ. When integrated by parts, defining $\hat{t} = -\frac{1}{2}\nabla^2$, $T = \int_\tau d^3\mathbf{r}\ \psi^*(\mathbf{r})\hat{t}\psi(\mathbf{r}) + \frac{1}{2}\int_\sigma \psi^*(\boldsymbol{\sigma})\nabla\psi(\boldsymbol{\sigma})\cdot d\boldsymbol{\sigma}$. With the usual boundary conditions for a large volume, the surface term vanishes, leaving the volume term as defined in subsequent literature, $T = (\psi|\hat{t}|\psi)$. The surface term becomes important in scattering theory and in solid-state theory using local atomic cells.

Defining the functional $V = (\psi|v|\psi)$, where $v(\mathbf{r})$ is a local potential for which $E = T + V$ is bounded below, the Schrödinger variational principle requires E to be stationary subject to normalization $(\psi|\psi) = 1$. The variation δE induced by $\delta\psi$ is expressed using the *functional derivative* $\frac{\delta E}{\delta\psi^*} = \mathcal{H}\psi$, defined such that $\delta E = \delta(\psi|\mathcal{H}|\psi) = \int d^3\mathbf{r}\{\delta\psi^* \frac{\delta E}{\delta\psi^*} + cc\}$. The variational condition is

$$\delta(E - \epsilon[(\psi|\psi) - 1]) = \int d^3\mathbf{r}\left(\delta\psi^*\left\{\frac{\delta E}{\delta\psi^*} - \epsilon\psi\right\} + cc\right) = 0,$$

for unconstrained $\delta\psi$ in the orbital Hilbert space, which implies the Schrödinger equation

$$\frac{\delta E}{\delta\psi^*} = \mathcal{H}\psi = \{\hat{t} + v\}\psi = \epsilon\psi.$$

Use of the Rayleigh quotient $\Lambda[\psi] = (\psi|\mathcal{H}|\psi)/(\psi|\psi)$ as a variational functional is an alternative to using a Lagrange multiplier to ensure normalization. Assuming that the eigenfunctions $\{\psi_i\}$ of $\mathcal{H}$ are complete and orthonormal in the relevant Hilbert space, $(\psi|\psi) = \sum_i |(i|\psi)|^2$ and $(\psi|\mathcal{H}|\psi) = \sum_i \epsilon_i|(i|\psi)|^2 \geq \epsilon_0(\psi|\psi)$. This implies that $\Lambda[\psi] \geq \epsilon_0$, a variational principle for the ground state. For other eigenstates, $\{\mathcal{H} - \epsilon_i\}\psi_i = 0$, so that $\Lambda[\phi_i] = \epsilon(i|i)/(i|i) = \epsilon_i$. Let $\psi = \psi_i + \delta\psi$. Then $\delta(\psi|\psi) = (\delta\psi|i) + (i|\delta\psi)$, and $\delta(\psi|\mathcal{H}|\psi) = (\delta\psi|\mathcal{H}|i) + (i|\mathcal{H}|\delta\psi) = \epsilon_i\delta(\psi|\psi)$. It then follows that

$$\delta\Lambda|_{\psi\to\psi_i} = \left.\frac{\delta(\psi|\mathcal{H}|\psi)(\psi|\psi) - (\psi|\mathcal{H}|\psi)\delta(\psi|\psi)}{(\psi|\psi)^2}\right|_{\psi\to\psi_i}$$
$$= \frac{\delta(\psi|\mathcal{H}|\psi) - \epsilon_i\delta(\psi|\psi)}{(\psi_i|\psi_i)} = 0.$$

By construction, this stationary principle holds for any eigenstate.

4.1.4 Basis set expansions

For expansion in a given set of basis functions that lie in the relevant Hilbert space, such that $\psi_i \simeq \sum_a \eta_a x_{ai}$, $(\psi|\mathcal{H}|\psi) = \sum_{a,b} x_a^*(a|\mathcal{H}|b)x_b$ is to be made stationary with respect to variations for which the normalization integral $(\psi|\psi) = \sum_{a,b} x_a^*(a|b)x_b$ is constant. For variations of the expansion coefficients,

$$\sum_{a,b} \delta x_a^*\{(a|\mathcal{H}|b) - \epsilon(a|b)\}x_b = 0,$$

which implies

$$\sum_b \{(a|\mathcal{H}|b) - \epsilon(a|b)\}x_b = 0.$$

In terms of the matrices $(a|\mathcal{H}|b)$ and $(a|b)$, the set of coefficients x_b is a null vector of the matrix $(a|\mathcal{H}|b) - \epsilon(a|b)$ when ϵ is determined so that the determinant of this matrix vanishes. If the basis functions are orthonormalized, this condition becomes $\det(h_{ab} - \epsilon\delta_{ab}) = 0$ and the coefficient vector is an eigenvector of the Hermitian matrix h_{ab}.

Suppose that $h^{(n-1)}$ is diagonalized in a basis of dimension $n-1$, and this basis is extended by adding an orthonormalized function η_n. The diagonalized matrix is augmented by a final row and column, with elements h_{ni}, h_{in} respectively, for $i < n$. The added diagonal element is h_{nn}. Modified eigenvalues are determined by the condition that the bordered determinant of the augmented matrix $h^{(n)} - \epsilon$ should vanish. This is expressed by

$$\det\left(h^{(n)} - \epsilon\right) = \Pi_{i<n}(\epsilon_i - \epsilon)\left[(h_{nn} - \epsilon) - \sum_{i<n}\frac{|h_{ni}|^2}{\epsilon_i - \epsilon}\right] = 0.$$

If the matrix elements h_{ni} do not vanish, the final factor may vanish at values of ϵ different from the original eigenvalues, giving the equation

$$h_{nn} - \epsilon = \sum_{i<n}\frac{|h_{ni}|^2}{\epsilon_i - \epsilon}.$$

The function on the right here has poles at each of the original eigenvalues. It has positive slope everywhere, because

$$\frac{d}{d\epsilon}\frac{|h_{ni}|^2}{\epsilon_i - \epsilon} = \frac{|h_{ni}|^2}{(\epsilon_i - \epsilon)^2}$$

for each term in the sum. This function increases from $-\infty$ to ∞ between each adjacent pair of unequal original eigenvalues, and must cross the straight line $h_{nn} - \epsilon$ once in each such interval.

Hence the new set of eigenvalues interleaves with the the original set. One new eigenvalue lies below all of the originals, one lies above, and the rest each occur in one of the intermediate intervals. This construction shows that the mth old eigenvalue is an upper bound for the mth new eigenvalue, for $m < n$. It follows that each of the n eigenvalues of a variational matrix eigenvalue problem is an upper bound for the corresponding eigenvalue of the Hamiltonian operator. If the eigenvalues are indexed in nondecreasing sequence, and h_{nn} falls between two adjacent nondegenerate eigenvalues ϵ_{k-1} and ϵ_k, the new eigenvalues $\epsilon_i^{(n)}$ for $i \neq k$ are displaced away from h_{nn}. Thus $\epsilon_i^{(n)} \leq \epsilon_i$ $(i < k)$, and $\epsilon_{i+1}^{(n)} \geq \epsilon_i$ $(i \geq k)$. This displacement rule is evident from the 2×2 submatrix for η_n and any ψ_i. The secular equation is $(h_{nn} - \epsilon_i^{(n)})(\epsilon_i - \epsilon_i^{(n)}) = |h_{ni}|^2$, which requires both factors to have the same sign.

4.2 Hellmann–Feynman and virial theorems

4.2.1 Generalized Hellmann–Feynman theorem

The derivation here follows Hurley [179]. Given $H(\xi)$ for some real parameter ξ, and a variational approximation such that $\delta(\Psi|H|\Psi) = 0$ and $\delta(\Psi|\Psi) = 0$, the following theorem can be proved: if variations $\delta\Psi$ include all variations induced by $\delta\xi$, then

$$\frac{\partial E}{\partial \xi} = \left\langle \frac{\partial H}{\partial \xi} \right\rangle.$$

The notation $\langle \cdots \rangle$ here denotes a mean value, $(\Psi|\cdots|\Psi)/(\Psi|\Psi)$. To prove the theorem, let each trial function depend on ξ, and require $\delta\Psi = \delta\xi \frac{\partial \Psi}{\partial \xi}$ to lie in the Hilbert space of variational trial functions. For variational wave functions and energy values, $(\delta\Psi|H - E|\Psi) + (\Psi|H - E|\delta\Psi) = 0$. For variations driven by $\delta\xi$,

$$\delta(\Psi|H - E|\Psi) = (\delta\Psi|H - E|\Psi) + (\Psi|H - E|\delta\Psi) \\ + \delta\xi \left(\Psi \left| \frac{\partial H}{\partial \xi} - \frac{\partial E}{\partial \xi} \right| \Psi \right) = 0$$

implies

$$\frac{\partial E}{\partial \xi} = \left(\Psi \left| \frac{\partial H}{\partial \xi} \right| \Psi \right) \Big/ (\Psi|\Psi) = \left\langle \frac{\partial H}{\partial \xi} \right\rangle,$$

which proves the theorem.

If the Hellmann–Feynman theorem is to be valid for forces on nuclei, the Coulomb cusp condition must be satisfied. However, if the nuclei are displaced, the orbital Hilbert space is modified. Hurley [179] noted this condition for finite basis sets, and introduced the idea of "floating" basis functions, with cusps that can shift away from the nuclei, in order to validate the theorem for such forces.

4.2.2 The hypervirial theorem

As an extension of Noether's theorem to quantum mechanics, the hypervirial theorem [101] derives conservation laws from invariant transformations of the theory. Consider a unitary transformation of the Schrödinger equation, $U(H - E)\Psi = U(H - E)U^\dagger U\Psi = 0$, and assume the variational Hilbert space closed under a unitary transformation $U(\xi) = e^{i\xi\hat{G}}$, where $\hat{G}$ is an Hermitian operator. Thus $i\hat{G}\Psi$

lies in the Hilbert space of variational trial functions. In the limit of an infinitesimal transformation, $\xi \to 0$, $(i\hat{G}\Psi|\Psi) + (\Psi|i\hat{G}\Psi) = 0$, and

$$(i\hat{G}\Psi|H - E|\Psi) + (\Psi|H - E|i\hat{G}\Psi) = 0$$
$$\Rightarrow (\Psi|[H, \hat{G}]|\Psi) = 0,$$

where $[H, \hat{G}] = H\hat{G} - \hat{G}H$. Since normalization is preserved, this implies the hypervirial theorem:

$$\langle [H, \hat{G}] \rangle = 0.$$

If $U(\xi) = e^{i\mathcal{G}(\xi)}$ such that $\mathcal{G}(0) = 0$, then

$$\left\langle \left[H, \frac{\partial \mathcal{G}}{\partial \xi} \right] \right\rangle \Bigg|_{\xi \to 0} = 0.$$

As an example, suppose for an N-electron system that energy E is approximated by an orbital functional $E[\{\phi_i, n_i\}]$, which depends on one-electron orbital wave functions ϕ_i and on occupation numbers n_i through a variational N-electron trial wave function Ψ. A momentum displacement is generated by $U = \exp(\frac{1}{i\hbar}\boldsymbol{\pi} \cdot \hat{\mathbf{D}})$, where $\hat{\mathbf{D}} = \sum_i \mathbf{r}_i$. In the momentum representation of the orbital wave functions, $U\Psi(\ldots, \phi_i(\mathbf{p}_i), \ldots) = \Psi(\ldots, \phi_i(\mathbf{p}_i + \boldsymbol{\pi}), \ldots)$. If the Hamiltonian is invariant under such a displacement, then $\frac{i}{\hbar}[H, \hat{\mathbf{D}}] = \hat{\mathbf{P}}$ and the hypervirial theorem implies conservation of total linear momentum $\langle \hat{\mathbf{P}} \rangle = \mathbf{0}$.

A coordinate displacement is generated by $U = \exp(\frac{i}{\hbar}\boldsymbol{\xi} \cdot \hat{\mathbf{P}})$ such that $U\Psi(\ldots, \phi_i(\mathbf{r}_i), \ldots) = \Psi(\ldots, \phi_i(\mathbf{r}_i + \boldsymbol{\xi}), \ldots)$. In this case, $\frac{i}{\hbar}[H, \hat{\mathbf{P}}] = -\sum_i \frac{\partial H}{\partial \mathbf{r}_i} = \hat{\mathbf{F}}$, the total force acting on the electrons. The hypervirial theorem implies that the mean value of this force vanishes for an isolated system, $\langle \hat{\mathbf{F}} \rangle = \mathbf{0}$.

4.2.3 The virial theorem

An N-electron virial operator is defined by

$$\mathcal{V} = \frac{1}{2} \sum_i (\mathbf{r}_i \cdot \mathbf{p}_i + \mathbf{p}_i \cdot \mathbf{r}_i) = \sum_i \mathbf{r}_i \cdot \mathbf{p}_i - 3i\hbar(N/2).$$

Using $(\frac{\hbar}{i}\frac{d}{d\xi} - \mathcal{V})e^{\frac{i}{\hbar}\xi\mathcal{V}}\Psi|_{\xi \to 0} = 0$, this operator generates a scale transformation $e^{\frac{i}{\hbar}\xi\mathcal{V}}\Psi(\ldots, \mathbf{r}_i, \ldots) = \tau^{3N/2}\Psi(\ldots, \tau\mathbf{r}_i, \ldots)$, where $\tau = e^{\xi}$. The commutator in the hypervirial theorem is

$$\frac{i}{\hbar}[H, \mathcal{V}] = \sum_i \left(\frac{\partial H}{\partial \mathbf{p}_i} \cdot \mathbf{p}_i - \mathbf{r}_i \cdot \frac{\partial H}{\partial \mathbf{r}_i} \right).$$

If the Hilbert space of the variational basis set is invariant under scale transformation, then the hypervirial theorem implies

$$\left\langle \sum_i \left(\frac{\partial H}{\partial \mathbf{p}_i} \cdot \mathbf{p}_i - \mathbf{r}_i \cdot \frac{\partial H}{\partial \mathbf{r}_i} \right) \right\rangle = 0. \tag{4.1}$$

In the nonrelativistic Hamiltonian, the kinetic energy T is homogeneous of degree 2 in the electronic momenta $\mathbf{p}_i$, and the Coulombic energy, including nuclear repulsions, is homogeneous of degree -1 in the electronic coordinates $\mathbf{r}_i$ and the nuclear coordinates $\mathbf{R}_a$. From Euler's theorem on partial derivatives of homogeneous functions,

$$\begin{aligned} \sum_i \frac{\partial H}{\partial \mathbf{p}_i} \cdot \mathbf{p}_i &= 2T, \\ \sum_i \frac{\partial H}{\partial \mathbf{r}_i} \cdot \mathbf{r}_i &= -V - \sum_a \frac{\partial H}{\partial \mathbf{R}_a} \cdot \mathbf{R}_a. \end{aligned} \tag{4.2}$$

The force acting on nucleus a is $\mathbf{F}_a = -\frac{\partial H}{\partial \mathbf{R}_a}$. From Eqs. (4.1) and (4.2), the general virial theorem for a molecule is

$$2\langle T \rangle + \langle V \rangle - \sum_a \langle \mathbf{F}_a \rangle \cdot \mathbf{R}_a = 0.$$

For a diatomic molecule with internuclear distance R,

$$2\langle T \rangle + \langle V \rangle + R \frac{\partial E}{\partial R} = 0.$$

4.3 The *N*-electron problem

4.3.1 The *N*-electron Hamiltonian

The N-electron wave function Ψ is an antisymmetric function of N sets of spatial and spin coordinates $\mathbf{r}_i$, s_i for individual electrons, all evaluated at a common time t. In postulating a time-dependent Schrödinger equation of the form

$$H\Psi = i\hbar \frac{\partial}{\partial t} \Psi,$$

the nonrelativistic N-electron theory is inherently inconsistent with the underlying relativistic field theory of electrons. Relativistic covariance in principle requires assigning a separate time variable t_i to each electron. The time-independent nonrelativistic theory of stationary states has a deeper justification, since assuming instantaneous interactions is not inconsistent with time-averaged interactions. The theory will be developed here primarily for stationary states. In this context, the N-electron Schrödinger equation is an eigenvalue equation in $4N$ spatial and spin coordinates,

$$(H - E)\Psi = 0,$$

where $H = \hat{T} + \hat{U} + \hat{V}$. The individual operators, for kinetic energy, interelectronic Coulomb interaction, and the external potential, respectively, are

$$\hat{T} = -\frac{1}{2}\sum_i \nabla_i^2,$$

$$\hat{U} = \frac{1}{2}\sum_{i,j}\frac{1}{r_{ij}} = \frac{1}{2}\sum_{i,j} u(r_{ij}),$$

$$\hat{V} = \sum_i v(\mathbf{r}_i).$$

Spin indices and summations will be suppressed in the notation here, but are to be understood in connection with coordinate indices and integrations over coordinates. Thus $\int d^3\mathbf{r}\cdots$ is intended to imply an implicit summation over a spin index. The energy functional for an antisymmetric wave function $\Psi(1, \ldots, N)$ is $(\Psi|H|\Psi) = \int_1 d^3\mathbf{r}_1 \cdots \int_N d^3\mathbf{r}_N \Psi^* H \Psi$ and the normalization integral is $(\Psi|\Psi) = \int_1 d^3\mathbf{r}_1 \cdots \int_N d^3\mathbf{r}_N \Psi^*\Psi$.

The Schrödinger variational principle requires $(\Psi|H|\Psi)$ to be stationary subject to constant normalization $(\Psi|\Psi)$. Introducing a Lagrange multiplier, the variational condition is

$$\delta(\Psi|H - E|\Psi) = (\delta\Psi|H - E|\Psi) + (\Psi|H - E|\delta\Psi) = 0,$$

for variations in the N-electron Hilbert space. The Hermitian character of H implies that $(\Psi|H - E|\delta\Psi) = (\delta\Psi|H - E|\Psi)^*$. Since $\delta\Psi$ is arbitrary, it can be multiplied by i. Then $(\Psi|H - E|\delta\Psi) = -(\delta\Psi|H - E|\Psi)^*$. Together, these equations imply that $(\Psi|H - E|\delta\Psi) = 0$ if and only if $(\delta\Psi|H - E|\Psi) = 0$. Hence if $\delta\Psi$ is unrestricted in the Hilbert space, the variational condition implies the time-independent Schrödinger equation $(H - E)\Psi = 0$.

4.3.2 Expansion in a basis of orbital wave functions

The N-electron Hamiltonian is a sum of one- and two-electron operators, $\hat{H} = \sum_i^N h(i) + \frac{1}{2}\sum_{i,j}^N u(i, j)$. A reference state Φ is defined as a single Slater determinant constructed as a normalized antisymmetrized product of N orthonormal orbital wave functions of the one-electron coordinates, $\Phi = \det\{\phi_1(1)\cdots\phi_N(N)\}$, implicitly including a normalization factor $(N!)^{-\frac{1}{2}}$. A complete set of such Slater determinants is determined by virtual excitations from this reference state, defined by replacing one or more of the N occupied orbital functions of Φ by functions from the residual unoccupied set. The general form is $\Phi_\mu = \det\{\phi_{1_\mu}(1)\cdots\phi_{N_\mu}(N)\}$. If the orbital functions are indexed by $i, j, \ldots \leq N$ for the occupied set, and by $N < a, b, \ldots$ for the residual unoccupied set, a consistent notation for virtual excitations is given by

$$\Phi_i^a = \det\{1, \ldots, a, \ldots, N\}, \qquad \Phi_{ij}^{ab} = \det\{1, \ldots, a, \ldots, b, \ldots, N\}, \ldots.$$

Taking spin indices into account, all two-electron integrals are of the form $(ij|\bar{u}|kl) = (ij|u|kl) - (ij|u|lk)$, with the convention that orbitals with different spin indices are orthogonal. It is convenient to truncate summations by the use of occupation numbers n_i, which are in principle determined by Fermi–Dirac statistics. At zero temperature, occupation numbers are determined by the structure of the reference state. Then $n_i = 1, n_a = 0$ for $i \leq N < a$. A convention used for double summation indices is $ij : i < j \leq N, ab : N < a < b$.

With these conventions of notation, simple formulas exist for all matrix elements of H in the basis of Slater determinants generated by virtual excitations from a reference state [72]. Denoting the latter by Φ_0, and defining an effective one-electron operator $\mathcal{H} = h + \sum_j n_j(ij|\bar{u}|ij)$ for Φ_0, the sequence of nondiagonal elements is

$$\begin{aligned}
\left({}_i^a|H|0\right) &= (a|\mathcal{H}|i),\\
\left({}_{ij}^{ab}|H|0\right) &= \sum_{ij} n_i n_j \sum_{ab}(1-n_a)(1-n_b)(ab|\bar{u}|ij),\\
\left({}_{ijk}^{abc}|H|0\right) &= 0, \text{etc.}
\end{aligned}$$

All such matrix elements vanish if more than two occupied orbitals are replaced, because H contains only a two-electron interaction operator. Given $(0|H|0) = \sum_i n_i(i|h|i) + \frac{1}{2}\sum_{i,j} n_i n_j(ij|\bar{u}|ij)$, diagonal elements follow a simple rule:

$$\begin{aligned}
\left({}_i^a|H|{}_i^a\right) - (0|H|0) &= (a|\mathcal{H}|a) - (i|\mathcal{H}|i) - (ai|\bar{u}|ai),\\
\left({}_{ij}^{ab}|H|{}_{ij}^{ab}\right) - (0|H|0) &= (a|\mathcal{H}|a) + (b|\mathcal{H}|b) - (i|\mathcal{H}|i) - (j|\mathcal{H}|j)\\
&\quad + (ab|\bar{u}|ab) + (ij|\bar{u}|ij) - (ai|\bar{u}|ai) - (bi|\bar{u}|bi)\\
&\quad - (aj|\bar{u}|aj) - (bj|\bar{u}|bj).
\end{aligned}$$

For $(\mu|H|\mu) - (0|H|0)$ defined by a virtual excitation $i, j, \ldots \to a, b, \ldots,$ the general rule is to add all elements $(a|\mathcal{H}|a)$, subtract all elements $(i|\mathcal{H}|i)$, add all elements of the form $(ab|\bar{u}|ab)$ or $(ij|\bar{u}|ij)$, and subtract all elements of the form $(ai|\bar{u}|ai)$.

4.3.3 The interelectronic Coulomb cusp condition

The electronic Coulomb interaction $u(r_{12}) = \frac{1}{r_{12}}$ greatly complicates the task of formulating and carrying out accurate computations of N-electron wave functions and their physical properties. Variational methods using fixed basis functions can only with great difficulty include functions expressed in relative coordinates. Unless such functions are present in a variational basis, there is an irreconcilable conflict with Coulomb cusp conditions at the singular points $r_{12} \to 0$ [23, 196]. No finite sum of product functions or Slater determinants can satisfy these conditions. Thus no practical restricted Hilbert space of variational trial functions has the correct structure of the true N-electron Hilbert space. The consequence is that the full effect of electronic interaction cannot be represented in simplified calculations.

The interelectronic Coulomb cusp can be analyzed by transforming a two-electron Hamiltonian to relative coordinates. The one-electron potential function is regular at the singularity $r_{12} \to 0$ and does not affect the cusp behavior. Given coordinates $\mathbf{r}_1$ and $\mathbf{r}_2$, mean and relative coordinates are defined, respectively, by

$$\mathbf{r} = \frac{1}{2}(\mathbf{r}_1 + \mathbf{r}_2),$$
$$\mathbf{q} = \mathbf{r}_2 - \mathbf{r}_1.$$

The Hamiltonian is

$$\begin{aligned} H &= -\frac{1}{2}\nabla_1^2 - \frac{1}{2}\nabla_2^2 + \frac{1}{r_{12}} + \mathcal{O}(1) \\ &= -\frac{1}{4}\nabla_r^2 - \nabla_q^2 + \frac{1}{q} + \mathcal{O}(1). \end{aligned}$$

Given relative angular momentum ℓ, the singular part of the Schrödinger equation is

$$\mathcal{H}_q = -\frac{\partial^2}{\partial q^2} - \frac{2}{q}\frac{\partial}{\partial q} + \frac{\ell(\ell+1)}{q^2} + \frac{1}{q} + \mathcal{O}(1).$$

A power-series solution is $f_\ell(q) = q^\ell(f_{\ell 0} + q f_{\ell 1} + \cdots)$, where $f_{\ell 0} \neq 0$ for $\ell \geq 0$. When substituted into the differential equation, this gives

$$\begin{aligned} \mathcal{H}_q f_\ell &= q^{\ell-2}\{[-\ell(\ell+1) + \ell(\ell+1)] f_{\ell 0}\} \\ &\quad + q^{\ell-1}\{[-(\ell+1)(\ell+2) + \ell(\ell+1)] f_{\ell 1} + f_{\ell 0}\} + \mathcal{O}(q^\ell). \end{aligned}$$

To avoid a singularity that cannot be cancelled by any one-electron potential, the coefficient of $q^{\ell-1}$ must vanish when $\ell = 0$. This implies the Coulomb cusp condition $f_0(q) = f_{00}(1 + \frac{1}{2}q + \cdots)$. A similar expansion is valid for any $\ell > 0$. Because the expansion includes odd powers of q, it describes a cusp in three dimensions, but implies a discontinuous gradient only for $\ell = 0$.

4.4 Symmetry-adapted functions

For systems with high symmetry, in particular for atoms, symmetry properties can be used to reduce the matrix of the N-electron Hamiltonian to separate noninteracting blocks characterized by global symmetry quantum numbers. A particular method will be outlined here [263], to complete the discussion of basis-set expansions. A symmetry-adapted function is defined by $\Theta = \mathcal{O}\Phi$, where $\mathcal{O}$ is an Hermitian projection operator ($\mathcal{O}^2 = \mathcal{O}$) that characterizes a particular irreducible representation of the symmetry group of the electronic Hamiltonian. Thus H commutes with $\mathcal{O}$. This implies the "turnover rule" $(\mathcal{O}\Phi|H|\mathcal{O}\Phi) = (\Phi|H|\mathcal{O}\Phi)$, which removes the projection operator from one side of the matrix element. Since the expansion of $\mathcal{O}\Phi$ may run to many individual terms, this can greatly simplify formulas and computing algorithms. Matrix elements $(\Theta_\mu|H|\Theta_\nu)$ simplify to $(\Phi_\mu|H|\Theta_\nu)$ or $(\Theta_\mu|H|\Phi_\nu)$.

For a general symmetry group, symmetry-adapted orbital functions can be indexed by $\nu\lambda\mu$, where ν designates a Hilbert space of dimension $d_{\nu\lambda}$. Each of the orthonormal basis orbitals belongs to an irreducible representation λ of the group, with subspecies index μ. Group elements are represented by matrices indexed by $\lambda\mu$ in this basis, which is closed under group operations. A *configuration* in the N-electron Hilbert space is characterized by a set of subshell occupation numbers $n_{\nu\lambda} \le d_{\nu\lambda}$ such that $\sum_\nu \sum_\lambda n_{\nu\lambda} = N$. Closed subshells are invariant and only open subshells ($n_{\nu\lambda} < d_{\nu\lambda}$) are relevant to the subsequent analysis. A configuration defines a Hilbert space whose basis is the set of Slater determinants defined by all independent choices of subspecies indices μ for the indicated open subshells. This set of basis determinants $\{\Phi_i, i = 1, n\}$ is closed under group operations and defines a group representation of $n \times n$ transformation matrices.

The representation of the N-electron symmetry group generated by a given configuration is block-diagonalized into irreducible representations indexed by Λ, with subspecies index M. In general there will be some number $m \le n$ of independent functions with the same symmetry indices. If the configuration basis is $\{\Phi_i, i = 1, n\}$, then a set of $m \le n$ unnormalized *projected determinants* $\Theta_j = \mathcal{O}\Phi_j = \ell_j^{-1} \sum_{i=1}^n y_{ji}\Phi_i$ are to be constructed, where $\mathcal{O}$ is a projector onto the symmetry species Λ, M. The function $\ell_j^{\frac{1}{2}}\Theta_j$ is normalized if $\ell_j = \sum_{i=1}^n |y_{ji}|^2$ and $y_{jj} = 1$. The projected determinants can be transformed into an orthonormal set $\Psi_\mu = k_\mu^{-\frac{1}{2}} \sum_{i=1}^n x_{\mu i}\Phi_i$, where $k_\mu = \sum_{i=1}^n |x_{\mu i}|^2$. Alternatively, as an expansion in the basis of projected determinants, $\Psi_\mu = k_\mu^{\frac{1}{2}} \sum_{j=1}^{m} a_{\mu j_1} \Theta_j$. Using the turnover rule, the implied overlap integrals are $(\Psi_\mu|\Psi_\nu) = \sum_{i=1}^{m} (k_\mu^{\frac{1}{2}} a_{\mu i}^*)(k_\nu^{-\frac{1}{2}} x_{\nu i}) = \delta_{\mu\nu}$, where $\sum_{i=1}^{m} a_{\mu i}^* x_{\nu i} = \delta_{\mu\nu}$.

In the basis of these orthonormal symmetry-adapted functions, matrix elements of the invariant Hamiltonian are given for two different configurations A and B by

$$\begin{aligned}(\Psi_{\mu A}|H|\Psi_{\nu B}) &= (k_\mu/k_\nu)^{\frac{1}{2}} \sum_{j=1}^{m_A} \sum_{i=1}^{n_B} a_{\mu j}^{A*} x_{\nu i}^{B} (\Phi_{jA}|H|\Phi_{iB}) \\ &= (k_\nu/k_\mu)^{\frac{1}{2}} \sum_{i=1}^{n_A} \sum_{j=1}^{m_B} x_{\mu i}^{A*} a_{\nu j}^{B} (\Phi_{iA}|H|\Phi_{jB}).\end{aligned} \tag{4.3}$$

As an option, the normalization constants can be eliminated using

$$(\Psi_{\mu A}|H|\Psi_{\nu B}) = \left\{\left[\sum_j^A \sum_i^B a_{\mu j}^{A*} x_{\nu i}^{B} (\Phi_{jA}|H|\Phi_{iB})\right]\left[\sum_i^A \sum_j^B x_{\mu i}^{A*} a_{\nu j}^{B} (\Phi_{iA}|H|\Phi_{jB})\right]\right\}^{\frac{1}{2}}. \tag{4.4}$$

4.4.1 Algorithm for constructing symmetry-adapted functions

In order to construct orthonormal projected determinants, the symmetry conditions should be expressed as a set of homogeneous linear equations. For example, the projector $\mathcal{O}$ for a finite group is a weighted sum of the representation matrices. This implies homogeneous matrix equations $\{\mathcal{O} - I\}\Theta = 0$. For a rotation group, with step-up and step-down operators $J_\pm$, functions obtained for $M = J$ by solving $J_+\Theta = 0$ determine those with $M < J$ by using the step-down operator J_-. The algorithm consists of the following steps [263]:

1. Reduce the equations by Gaussian elimination from right to left.
2. Use back-substitution to determine the solution whose leading coefficient corresponds to the rightmost undetermined pivot element. Set this coefficient to unity.
3. Append this solution vector to the set of equations and remove elements to the right of the new pivot element by Gauss elimination.
4. Repeat to obtain a trapezoidal array of m orthogonal null vectors $\{x_{\mu i}\}$, each with leading coefficient unity.
5. The last line is the projection of the first basis function.
6. Invert the leading triangle $x_{\mu i}$ to get $a_{\mu j}$.

Table 4.1. *a/x table for functions ^2S (abc)*

Equations			
$i = 2$	3	4	
1	−1	1	
Coefficients			k_μ
$-\frac{1}{2}$	1*	1	2
1*	$\frac{1}{2}$	$-\frac{1}{2}$	$\frac{3}{2}$

4.4.2 Example of the method

Consider the total spin eigenstates constructed from the three-electron configuration abc, consisting of all eight Slater determinants obtained by assigning spin α : $m_s = \frac{1}{2}$ or spin β : $m_s = -\frac{1}{2}$ to each of three orthonormal spatial orbital functions $\phi_a(\mathbf{r})$, $\phi_b(\mathbf{r})$, $\phi_c(\mathbf{r})$. Because total S_z is quantized, the full configuration basis can be separated into subsets indexed by M_S. Thus for $M_S = \frac{3}{2}$ there is only one basis function $\Phi_1 = \det(a_\alpha b_\alpha c_\alpha)$. The algorithm determines a single quartet ($S = \frac{3}{2}$) spin eigenfunction, for which $\Psi_1 = \Theta_1 = \Phi_1$ with coefficient vectors $a_{11} = x_{11} = 1$, and normalization constant $k_1 = 1$.

For $M_S = \frac{1}{2}$, there are three basis functions,

$$\Phi_2 = \det(a_\alpha b_\alpha c_\beta),$$

$$\Phi_3 = \det(a_\alpha c_\alpha b_\beta),$$

$$\Phi_4 = \det(b_\alpha c_\alpha a_\beta).$$

Coefficients $(\Phi_1|S_+|\Phi_i)$, $i = 2, 3, 4$ in the linear equation $S_+\Theta = 0$ are shown in Table (4.1), together with the vectors $x_{\mu i}$ generated using the present algorithm. Nondiagonal elements of the upper left-hand triangle of $x_{\mu i}$, all of which vanish by construction, are replaced in such a/x tables by elements of the triangular matrix $a_{\mu i}$. Diagonal elements common to both matrices a and x are marked by an asterisk.

If the one-electron and Coulomb matrix elements, which are the same for all basis functions of the configuration, are denoted by E_0, matrix elements of the Hamiltonian in the $M_S = \frac{1}{2}$ subconfiguration are

$$H_{22} = E_0 - (ab|u|ba),$$

$$H_{23} = (bc|u|cb),$$

$$H_{33} = E_0 - (ac|u|ca),$$
$$H_{24} = -(ac|u|ca),$$
$$H_{34} = (ab|u|ba),$$
$$H_{44} = E_0 - (bc|u|cb).$$

Matrix elements for the two orthonormal projected doublet ($S = \frac{1}{2}$) functions, computed using Eq. (4.4), are

$$
\begin{aligned}
(\Psi_2|H|\Psi_2) &= H_{22} + \frac{1}{2}H_{23} - \frac{1}{2}H_{24} \\
&= E_0 - (ab|u|ba) + \frac{1}{2}(ac|u|ca) + \frac{1}{2}(bc|u|cb), \\
(\Psi_3|H|\Psi_2) &= \left\{\left[-\frac{1}{2}\left(H_{22} + \frac{1}{2}H_{23} - \frac{1}{2}H_{24}\right)\right.\right. \\
&\quad \left.\left. + \left(H_{32} + \frac{1}{2}H_{33} - \frac{1}{2}H_{34}\right)\right][(H_{23} + H_{24})]\right\}^{\frac{1}{2}} \\
&= \left\{\frac{3}{4}[-(ac|u|ca) + (bc|u|cb)][-(ac|u|ca) + (bc|u|cb)]\right\}^{\frac{1}{2}} \\
&= -\left(\frac{3}{4}\right)^{\frac{1}{2}}[(ac|u|ca) - (bc|u|cb)], \\
(\Psi_3|H|\Psi_3) &= -\frac{1}{2}(H_{23} + H_{24}) + (H_{33} + H_{34}) \\
&= E_0 + (ab|u|ba) - \frac{1}{2}(ac|u|ca) - \frac{1}{2}(bc|u|cb).
\end{aligned}
$$

5

Independent-electron models

The principal references for this chapter are:

[26] Blanchard, P. and Brüning, E. (1992). *Variational Methods in Mathematical Physics: A Unified Approach* (Springer-Verlag, Berlin).

[90] Dreizler, R.M. and Gross, E.K.U. (1990). *Density Functional Theory* (Springer-Verlag, Berlin).

[102] Eschrig, H. (1996). *The Fundamentals of Density Functional Theory* (Teubner, Stuttgart).

[130] Froese Fischer, C. (1977). *The Hartree-Fock Method for Atoms* (Wiley, New York).

[185] Janak, J.F. (1978). Proof that $\partial E/\partial n_i = \epsilon_i$ in density-functional theory, *Phys. Rev.* B**18**, 7165–7168.

[189] Jones, R.O. and Gunnarsson, O. (1989). The density functional formalism, its applications and prospects, *Rev. Mod. Phys.* **61**, 689–746.

[211] Landau, L.D. (1956). The theory of a Fermi liquid, *Zh. Eksp. Teor. Fiz.* **30**, 1058–1064. [*Sov. Phys. JETP* **3**, 920–925 (1956)]

[212] Landau, L.D. (1957). Oscillations in a Fermi liquid, *Zh. Eksp. Teor. Fiz.* **32**, 59–66. [*Sov. Phys. JETP* **5**, 101–108 (1957)]

[239] McWeeny, R. (1989). *Methods of Molecular Quantum Mechanics* (Academic Press, New York).

[321] Parr, R.G. and Yang, W. (1989). *Density-Functional Theory of Atoms and Molecules* (Oxford University Press, New York).

[386] Slater, J.C. (1972). Statistical exchange-correlation in the self-consistent field, *Adv. Quantum Chem.* **6**, 1–92.

[388] Slater, J.C. and Wood, J.H. (1971). Statistical exchange and the total energy of a crystal, *Int. J. Quantum Chem. Suppl.* **4**, 3–34.

Despite the simple and universal structure of the nonrelativistic Hamiltonian for N interacting electrons, it produces a broad spectrum of physical and chemical phenomena that are difficult to conceptualize within the full N-electron theory. Starting with the work of Hartree [162] in the early years of quantum mechanics, it was found to be very rewarding to develop a model of electrons that interact only indirectly with each other, through a *self-consistent* mean field. A deeper motivation lies in the fact that the relativistic quantum field theory of electrons is explicitly described by a field operator that corresponds more closely to a one-particle model wave function than to that of the Schrödinger N-electron theory. The fundamental characterization of this electron field by Fermi–Dirac statistics is directly applicable to the mean-field theory, using concepts of statistical occupation numbers determined by effective one-electron orbital energy values. The variational theory appropriate to such independent-electron models is developed in this chapter.

Hartree's original idea of the self-consistent field involved only the direct Coulomb interaction between electrons. This is not inconsistent with variational theory [163], but requires an essential modification in order to correspond to the true physics of electrons. In neglecting electronic exchange, the pure Coulombic Hartree mean field inherently allowed an electron to interact with itself, one of the most unsatisfactory aspects of pre-quantum theories. Hartree simply removed the self-interaction by fiat, at the cost of making the mean field different for each electron. Orbital orthogonality, necessary to the concept of independent electrons, could only be imposed by an artificial variational constraint. The need for an *ad hoc* self-interaction correction (SIC) persists in recent theories based on approximate local exchange potentials.

In a theory using antisymmetric wave functions [127, 79] the interaction of an electron with itself disappears due to cancellation between classical Coulombic and exchange interactions. The Hartree–Fock theory, including exact exchange, became an essential methodology in atomic and molecular physics. Computational procedures and results for atoms have been reviewed by Hartree [163] and by Froese Fischer [130]. The electronic *correlation* energy, neglected in the Hartree–Fock approximation, is important for an exact description of electronic phenomena in atoms, molecules, and solids, but is difficult to treat in practical computational theory. Before discussing Hartree–Fock theory and other mean-field approximations here, a formally exact independent-electron theory will be developed. This *orbital functional theory* (OFT) embodies the correlation theory of Brueckner [43], originally a rationalization of the nuclear shell model, which becomes an exact theory when applied to the explicit two-electron Coulomb interaction.

5.1 N-electron formalism using a reference state

Although a Slater-determinant reference state Φ cannot describe such electronic "correlation" effects as the wave-function modification required by the interelectronic Coulomb singularity, a variationally based choice of an optimal reference state can greatly simplify the N-electron formalism. Φ defines an orthonormal set of N occupied orbital functions ϕ_i with occupation numbers $n_i = 1$. While $(\Phi|\Phi) = 1$ by construction, for any full N-electron wave function Ψ that is to be modelled by Φ it is convenient to adjust $(\Psi|\Psi) \geq 1$ to the unsymmetrical normalization $(\Phi|\Psi) = 1$. This implies that $(\Phi|\Psi - \Phi) = 0$, so that Ψ is partitioned into complementary orthogonal components Φ and $\Psi - \Phi$. Because $E\Psi = H\Psi$ for any energy eigenstate, $E = (\Phi|H|\Psi) = E_0 + E_c$, where $E_0 = (\Phi|H|\Phi)$ is the mean energy of the reference state, while $E_c = (\Phi|H|\Psi - \Phi)$ provides a natural definition of correlation energy. Thus any rule $\Psi \to \Phi$ that determines a reference state also defines the correlation energy.

Concerned with reconciling the strong internucleon interaction, described empirically by a hard-core interaction, with the success of the "independent-particle" nuclear shell model, Brueckner [43] proposed that for any given Ψ, with arbitrary normalization, an optimal reference state could be defined by maximizing the projection $(\Phi|\Psi)$. Since $H\Psi = E\Psi$ for an energy eigenstate, the Brueckner variational condition $(\delta\Phi|\Psi) = 0$ implies that $(\delta\Phi|H|\Psi) = 0$. Following an argument used to derive Hartree–Fock theory, variations that just mix occupied orbitals do not affect the antisymmetrized function Φ. Meaningful infinitesimal variations $\delta\Phi$ can be expressed as a sum of single virtual excitations Φ_i^a. The corresponding variational condition in Brueckner theory can be expressed in terms of an effective one-electron "Hamiltonian" $\mathcal{G}$ defined such that $(\Phi_i^a|H|\Psi) = (a|\mathcal{G}|i) = 0$ for $i \leq N < a$. This heuristic derivation from Brueckner's variational condition [275] does not fully determine the operator $\mathcal{G}$, and explicitly assumes that Ψ is given in advance. In any model theory that expresses both E_0 and E_c as *orbital functionals*, depending only on occupied orbitals of a reference state Φ and on occupation numbers n_i, the linear operator $\mathcal{G}$ is determined by the orbital functional derivative $\frac{\delta E}{n_i \delta\phi_i^*} = \mathcal{G}\phi_i$. Introducing Lagrange multipliers for orthonormality, the variational condition $\delta\{E - \sum_{i,j}[(\phi_i|\phi_j) - \delta_{ij}]\lambda_{ji}\} = 0$ implies the orbital Euler–Lagrange (OEL) equations of orbital functional theory,

$$\{\mathcal{G} - \epsilon_i\}\phi_i = 0,$$

for the occupied orbitals of the reference state in a diagonalized representation of the matrix λ_{ji}. An iterative "self-consistent" procedure is required in general, because both E_0 and E_c depend nonlinearly on the occupied orbital set. When $\mathcal{G}$ is Hermitian, $\mathcal{G}\phi_i = \phi_i\epsilon_i$ and $\phi_i^*\mathcal{G} = \epsilon_i\phi_i^*$. Defining the Dirac density matrix $\hat{\rho}$, whose kernel is $\sum_i \phi_i(\mathbf{r})n_i\phi^*(\mathbf{r}')$, this implies $[\mathcal{G}, \hat{\rho}] = 0$ as an alternative statement of the OEL equations.

5.1.1 Fractional occupation numbers

For applications to open-shell states of atoms and molecules or to metallic solids, and for systems at finite temperature, it is convenient to treat occupation numbers as parameters that can vary freely in the range $0 \leq n_i \leq 1$. If the occupation numbers

$\{n_i\}$ in orbital functional theory are treated as continuous variables, the extended energy functional $E[\{\phi, n_i\}]$ provides a natural interpolation between quantum states characterized by an explicit reference state Φ. Formal theory based on this interpolation will be considered here, setting aside the somewhat controversial issue of physical realization of such interpolations. Landau [211, 212] introduced this concept as appropriate to a model of fermionic *quasiparticles* analogous to the dressed electrons of renormalized quantum electrodynamics. The theory is characterized by empirical expressions for total energy and entropy that are parametric functions of orbital occupation numbers. The orbital functional theory developed below, especially in its time-dependent extension, is a particular realization of Landau's concept. Energy relationships are determined by Janak's theorem, $\epsilon_i = \frac{\partial E}{\partial n_i}$, which will be proven below for general OFT models.

Orbital functional theory defines energy functionals for fractional occupation numbers, providing a smooth interpolation between the values 0, 1 appropriate to specific wave functions. Following Slater and Wood [388], changes such as addition or removal of an electron can be described by integrating fractional occupation numbers between the appropriate limits, using derivatives $\partial E/\partial n_i$ derived from solution of OEL equations with intermediate fractional occupation numbers. An excitation energy can be estimated by a single calculation of the "transition state" defined with occupation numbers averaged over the initial and final states [386].

5.1.2 Janak's theorem

When the variational energy is a functional of the reference state Φ, it can always be expressed in terms of integrals over the Dirac matrix $\sum_i \phi_i(\mathbf{r}) n_i \phi_i^*(\mathbf{r}')$, which remains well-defined for nonintegral occupation numbers. This implies two chain rules, for $n_i \neq 0$:

$$\frac{\delta E}{n_i \delta \phi_i^*} = \mathcal{G}\phi_i = \frac{\delta E}{\delta(n_i \phi_i^*)},$$
$$\frac{\partial E}{\partial n_i} = \int d^3\mathbf{r}\, \phi_i^* \frac{\delta E}{\delta(n_i \phi_i^*)} = \int d^3\mathbf{r}\, \phi_i^* \mathcal{G}\phi_i.$$

Hence if n_i is allowed to vary,

$$\delta E = \sum_i \delta n_i \int d^3\mathbf{r}\, \phi_i^* \mathcal{G}\phi_i + \sum_i n_i \int d^3\mathbf{r} \{\delta\phi_i^* \mathcal{G}\phi_i + cc\},$$

where $\int \phi_i^* \mathcal{G}\phi_i = \epsilon_i$ and $\int \delta\phi_i^* \mathcal{G}\phi_i = \epsilon_i \int \delta\phi_i^* \phi_i = 0$ for normalized occupied orbital solutions of the OEL equations. This implies that a theorem proved by Janak [185] is generally valid in OFT. For variations of occupation numbers in which occupied orbitals are "relaxed" to satisfy the OEL equations, Janak's theorem takes the form

$$dE = \sum_i \frac{\partial E}{\partial n_i} dn_i = \sum_i \epsilon_i \, dn_i.$$

This theorem justifies the procedure of Slater and Wood for any version of orbital functional theory.

5.2 Orbital functional theory

5.2.1 Explicit components of the energy functional

$E_0 = (\Phi|H|\Phi)$ is a sum of explicit orbital functionals,

$$T = (\Phi|\hat{T}|\Phi) = \sum_i n_i (i|\hat{t}|i),$$

$$U = (\Phi|\hat{U}|\Phi) = \frac{1}{2}\sum_{i,j} n_i n_j (ij|\bar{u}|ij),$$

$$V = (\Phi|\hat{V}|\Phi) = \sum_i n_i (i|v|i),$$

where $\hat{t} = -\frac{1}{2}\nabla^2$ and $\bar{u} = \frac{1}{r_{12}}(1 - \mathcal{P}_{12})$ incorporates an operator $\mathcal{P}_{12}$ that exchanges space-spin coordinates of two electrons. This defines Coulomb and exchange functionals such that $(\Phi|\hat{U}|\Phi) = E_h + E_x$, where

$$E_h = \frac{1}{2}\sum_{i,j} n_i n_j (ij|u|ij),$$

$$E_x = -\frac{1}{2}\sum_{i,j} n_i n_j (ij|u|ji).$$

The corresponding orbital functional derivatives are defined by

$$\frac{\delta T}{n_i \delta \phi_i^*} = \hat{t}\phi_i,$$

$$\frac{\delta V}{n_i \delta \phi_i^*} = v(\mathbf{r})\phi_i,$$

$$\frac{\delta E_h}{n_i \delta \phi_i^*} = v_h(\mathbf{r})\phi_i = \sum_j n_j (j|u|j)\phi_i,$$

$$\frac{\delta E_x}{n_i \delta \phi_i^*} = \hat{v}_x \phi_i = -\sum_j n_j (j|u|i)\phi_j.$$

The correlation energy E_c, defined in general by $\langle H\rangle - (\Phi|H|\Phi)$, takes the particular form $E_c = (\Phi|H|\mathcal{Q}\Psi)$, using the unsymmetrical normalization $(\Phi|\Psi) = (\Phi|\Phi) = 1$ and defining the projection operator $\mathcal{Q} = I - \Phi\Phi^\dagger$, determined by Φ. In any practical theoretical model, E_c must be approximated, and the functional derivative $\delta E_c/n_i\delta\phi_i^* = \hat{v}_c\phi_i$ is to be evaluated from this approximate expression.

5.2.2 Orbital Euler–Lagrange equations

The reference state Φ is a Slater determinant constructed as a normalized antisymmetrized product of N orthonormal spin-indexed orbital functions $\phi_i(\mathbf{r})$. The orbital energy functional $E = E_0 + E_c$ is to be made stationary, subject to the orbital orthonormality constraint $(i|j) = \delta_{ij}$, imposed by introducing a matrix of Lagrange multipliers λ_{ji}. The variational condition is

$$\delta\left\{E - \sum_{ij} n_i n_j \left(\int \phi_i^*\phi_j d^3\mathbf{r} - \delta_{ij}\right)\lambda_{ji}\right\} =$$
$$\sum_i n_i \left[\int \delta\phi_i^* \left\{\frac{\delta E}{n_i\delta\phi_i^*} - \sum_j n_j\phi_j\lambda_{ji}\right\} d^3\mathbf{r} + cc\right]$$
$$+ \sum_i \delta n_i \int \phi_i^* \frac{\delta E}{n_i\delta\phi_i^*} d^3\mathbf{r} = 0. \tag{5.1}$$

With fixed $n_i \neq 0$, for variations of occupied orbitals that are unconstrained in the orbital Hilbert space, the variational condition implies orbital Euler–Lagrange equations

$$\frac{\delta E}{n_i\delta\phi_i^*} = \mathcal{G}\phi_i = \sum_j n_j\phi_j\lambda_{ji}. \tag{5.2}$$

The orbital functional derivative here defines an effective Hamiltonian

$$\mathcal{G} = -\frac{1}{2}\nabla^2 + v(\mathbf{r}) + v_h(\mathbf{r}) + \hat{v}_x + \hat{v}_c.$$

The theory is usually expressed in terms of *canonical* equations

$$\{\mathcal{G} - \epsilon_i\}\phi_i = 0,$$

obtained by diagonalizing the matrix of Lagrange multipliers for the occupied orbitals.

5.2.3 Exact correlation energy

It was shown by Brenig [35] that $(\delta\Phi|\Psi) = 0$ removes all particle–hole virtual excitations Φ_i^a from the wave function Ψ. This follows because $(\delta\Phi|\Psi) = \sum_{i,a} \delta c_i^{a*}(\Phi_i^a|\Psi)$ implies $(\Phi_i^a|\Psi) = 0$ for $i \leq N < a$. Since there are only two-particle interactions in the N-electron Hamiltonian, an exact formula for the correlation energy can be expressed with only two-electron matrix elements:

$$E_c = (\Phi|H|\Psi - \Phi) = \sum_{ij}\sum_{ab}(0|H|_{ij}^{ab})c_{ij}^{ab} = \sum_{ij}\sum_{ab}(ij|\bar{u}|ab)c_{ij}^{ab}.$$

By construction $|c_{ij}^{ab}| = |(\Phi_{ij}^{ab}|\Psi)| \leq (\Phi|\Psi) = 1$. c_{ij}^{ab} can be considered to define a two-electron operator with antisymmetric matrix elements $(ab|\bar{c}|ij)$, such that

$$E_c = \sum_{ij} n_i n_j \sum_{ab}(1 - n_a)(1 - n_b)(ij|\bar{u}|ab)(ab|\bar{c}|ij).$$

The coefficients c_{ij}^{ab} can be obtained only by constructing the wave function Ψ. An exact formal expression can be derived by partitioning the N-electron Hilbert space using the projection operator $\mathcal{P} = \Phi\Phi^\dagger$ and its orthogonal complement $\mathcal{Q} = I - \Phi\Phi^\dagger$. Using the definition given above, the correlation energy is $E_c = (\Phi|H|\mathcal{Q}\Psi)$. A more explicit expression is obtained from the partitioned Schrödinger equation

$$\mathcal{P}(H - E)\mathcal{P}\Psi + \mathcal{P}(H - E)\mathcal{Q}\Psi = 0,$$
$$\mathcal{Q}(H - E)\mathcal{P}\Psi + \mathcal{Q}(H - E)\mathcal{Q}\Psi = 0.$$

Here $\mathcal{Q}H\mathcal{Q}$ defines a reduced Hamiltonian operator $\hat{H}$. On substituting a formal solution of the second set of equations into the first, it follows for $\eta \to 0+$ that

$$E = (\Phi|H - H(\hat{H} - E - i\eta)^{-1}H|\Phi),$$

where $(\hat{H} - E - i\eta)^{-1}$ denotes an inverse operator in the $\mathcal{Q}$-space. This expresses the correlation energy as an exact implicit functional of Φ and hence of its occupied orbital functions [290],

$$E_c = -(\Phi|H(\hat{H} - E_0 - E_c - i\eta)^{-1}H|\Phi). \tag{5.3}$$

The coefficients c_{ij}^{ab} are given by an expression that depends only on the reference state and on a parametric value of E_c,

$$c_{ij}^{ab} = \left(\Phi_{ij}^{ab}|\mathcal{Q}\Psi\right)$$
$$= -\left(\Phi_{ij}^{ab}|(\hat{H} - E_0 - E_c - i\eta)^{-1}H|\Phi\right).$$

Since H is specified, Eq. (5.3) defines E_c as a functional of the occupied orbitals of Φ, although it cannot be expressed as an explicit closed formula. The orbital functional derivative $\frac{\delta E_c}{n_i \delta\phi_i^*} = \hat{v}_c\phi_i$ defines a nonlocal correlation potential $\hat{v}_c$ in exact OEL equations. Iterative solution for E_c combines a search for a root of the partitioned N-electron secular determinant with the Brueckner condition that defines Φ. This would be a legitimate variational method if the inverse of matrix $\mathcal{Q}(H-E)\mathcal{Q}$ could be evaluated exactly. In practice this is not possible, but standard approximations of many-body theory are available for particular applications of the formalism.

An infinitesimal unitary transformation of the orbital basis that modifies Φ must mix occupied and unoccupied orbital functions. For a typical orbital variation, $\delta\Phi = \Phi_i^a \delta c_i^a$ and $\delta\phi_i = \phi_a \delta c_i^a$. Unitarity induces $\delta\phi_a = -\phi_i \delta c_i^{a*} = -\phi_i(\delta\phi_i|\phi_a)$. In terms of functional derivatives $\delta/\tilde{\delta}\phi$ defined for independent variations of occupied and unoccupied orbitals,

$$\delta E_c = \sum_i n_i \left(\delta\phi_i \left| \frac{\delta E_c}{n_i \tilde{\delta}\phi_i^*}\right.\right) + \sum_a (1-n_a)\left(\left.\frac{\delta E_c}{(1-n_a)\tilde{\delta}\phi_a}\right| \delta\phi_a\right) + cc. \tag{5.4}$$

For such variations,

$$\delta E_c = \sum_i n_i \left(\delta\phi_i \left| \frac{\delta E_c}{n_i \tilde{\delta}\phi_i^*}\right.\right) - \sum_i n_i \sum_a (1-n_a)\left(\left.\frac{\delta E_c}{(1-n_a)\tilde{\delta}\phi_a}\right| \phi_i\right)(\delta\phi_i|\phi_a) + cc. \tag{5.5}$$

Treating the coefficients c_{ij}^{ab} as constants in the current cycle of an iteration loop, and using Eq. (5.5), $(a|\delta E_c/n_i\delta\phi_i^*)$ defines the matrix element

$$(a|\hat{v}_c|i) = \frac{1}{2}\sum_j n_j \sum_{b,c}(1-n_c)(1-n_b)(aj|\bar{u}|cb)(cb|\bar{c}|ij) - \frac{1}{2}\sum_{j,k} n_k n_j \sum_b (1-n_b)(kj|\bar{u}|ib)(ab|\bar{c}|kj), \tag{5.6}$$

in agreement with matrix elements deduced from the double virtual excitation terms in $\mathcal{Q}\Psi$ [275]. The effective correlation "potential" is a linear operator $\hat{v}_c$ with the kernel

$$v_c(\mathbf{r},\mathbf{r}') = \frac{1}{2}\sum_j n_j \sum_{b,c}(1-n_c)(1-n_b)(j|\bar{u}|b)\phi_c(\mathbf{r})\phi_c^*(\mathbf{r}')(b|\bar{c}|j) - \frac{1}{2}\sum_{j,k} n_k n_j \sum_b (1-n_b)(b|\bar{c}|j)\phi_k(\mathbf{r})\phi_k^*(\mathbf{r}')(j|\bar{u}|b). \tag{5.7}$$

This operator is consistent with the leading terms in quasiparticle self-energies implied by many-body theory [275, 407].

5.3 Hartree–Fock theory

5.3.1 Closed shells – unrestricted Hartree–Fock (UHF)

The theory is based on an optimized reference state Φ that is a Slater determinant constructed as a normalized antisymmetrized product of N orthonormal spin-indexed orbital functions $\phi_i(\mathbf{r})$. This is the simplest form of the more general orbital functional theory (OFT) for an N-electron system. The energy functional $E = (\Phi|H|\Phi)$ is required to be stationary, subject to the orbital orthonormality constraint $(i|j) = \delta_{ij}$, imposed by introducing a matrix of Lagrange multipliers λ_{ji}. The general OEL equations derived above reduce to the UHF equations if correlation energy E_c and the implied correlation potential $\hat{v}_c$ are omitted. The effective Hamiltonian operator is

$$\mathcal{H} = -\frac{1}{2}\nabla^2 + v(\mathbf{r}) + v_h(\mathbf{r}) + \hat{v}_x.$$

The theory is usually expressed in terms of *canonical* Hartree–Fock equations

$$\{\mathcal{H} - \epsilon_i\}\phi_i = 0,$$

obtained by diagonalizing the matrix of Lagrange multipliers for the occupied orbitals.

Janak's theorem, valid for general OEL equations when occupation numbers are varied, holds for the UHF theory in the form

$$\frac{\partial E}{\partial n_i} = \int \phi_i^* \mathcal{H} \phi_i \, d^3\mathbf{r} = \epsilon_i,$$

for infinitesimal changes of occupation numbers for which the occupied orbitals are relaxed as $\mathcal{H}$ changes. If orbital relaxation is neglected, the change of energy due to removing an occupied orbital ϕ_i is given by the general rule for diagonal elements cited in Section 4.3.2:

$$({}_i|H|{}_i) - (0|H|0) = -(i|\mathcal{H}|i) = -\epsilon_i,$$

which is a statement of Koopmans' theorem [206] for ionization potentials. The corresponding expression for electron affinities,

$$({}^a|H|{}^a) - (0|H|0) = (a|\mathcal{H}|a) = \epsilon_a,$$

must be used with caution, since typical variational calculations do not correctly represent the continuum of electronic states outside a neutral atom or molecule.

This subject will be treated in a later chapter on continuum states and scattering theory.

5.3.2 Brillouin's theorem

Because of antisymmetry, variations of Φ that are simply linear transformations of occupied orbitals have no effect other than a change of normalization. For orbital functions with fixed normalization, a general variation of Φ takes the form $\delta\Phi = \sum_i n_i \sum_a (1 - n_a)\Phi_i^a \delta c_i^a$. The variational condition is

$$\begin{aligned}\delta(\Phi|H|\Phi) &= \sum_i n_i \sum_a (1 - n_a)\{\delta c_i^{a*}({}_i^a|H|0) + cc\} \\ &= \sum_i n_i \sum_a (1 - n_a)\{\delta c_i^{a*}(a|\mathcal{H}|i) + cc\} = 0.\end{aligned}$$

This is a statement of Brillouin's theorem [37], that $(a|\mathcal{H}|i) = 0$, $i \le N < a$ is a necessary condition for $(\Phi|H|\Phi)$ to be stationary. The normalization of occupied variables must also be varied in order to determine the Lagrange multipliers ϵ_i. Definition of the effective Hamiltonian $\mathcal{H}$ requires diagonal matrix elements determined by $\delta E / n_i \delta\phi_i^*$ for unconstrained variations $\delta\phi_i$.

5.3.3 Open-shell Hartree–Fock theory (RHF)

The UHF formalism becomes inconvenient for open-shell configurations of atoms or molecules with point-group symmetry. Unless specific restrictions are imposed, the self-consistent occupied orbitals fall into sets that are nearly but not quite transformable into each other by operations of the symmetry group. By imposing "equivalence" and "symmetry" restrictions, these sets become symmetry-adapted basis states for irreducible representations of the symmetry group. This makes it possible to construct symmetry-adapted N-electron functions, as described in Section 4.4. The constraints in general invalidate the theorems of Brillouin and Koopmans. This "restricted" theory (RHF) is described in detail for atoms by Hartree [163] and by Froese Fischer [130].

To illustrate the modifications of UHF formalism, it is convenient to consider pure spin symmetry for a single Slater determinant with N_c doubly occupied spatial orbitals χ_i^c and N_o singly occupied orbitals χ_i^o. The corresponding UHF state has N_α : $m_s = \frac{1}{2}$ occupied spin orbitals ϕ_i^α and N_β : $m_s = -\frac{1}{2}$ occupied spin orbitals ϕ_i^β. The number of open-shell and closed-shell orbitals are, respectively $N_o = N_\alpha - N_\beta > 0$ and $N_c = N_\beta$. Occupation numbers for the spatial orbitals are $n^c = 2$, $n^o = 1$. If all orbital functions are normalized, a canonical form of the RHF reference state is defined by orthogonalizing the closed- and open-shell sets separately.

Nonzero Lagrange multipliers are required for orthogonalization of orbitals χ^o to χ^c. The energy functional $(\Phi|H|\Phi)$ is

$$\sum_i n_i(i|h|i) + \frac{1}{2}\sum_{i,j} n_i n_j (ij|u|ij) - \frac{1}{2}\left\{\frac{1}{2}\sum_{i,j}^{c,c} + \sum_{i,j}^{o,c} + \sum_{i,j}^{o,o}\right\} n_i n_j (ij|u|ji),$$

expressed in terms of the spatial orbitals. Alternatively,

$$(\Phi|H|\Phi) = \sum_i^c 2(i|h|i) + \sum_i^o (i|h|i) + \frac{1}{2}\sum_{i,j}^{c,c}[4(ij|u|ij) - 2(ij|u|ji)]$$
$$+ \sum_{i,j}^{o,c}[2(ij|u|ij) - (ij|u|ji)] + \frac{1}{2}\sum_{i,j}^{o,o}[(ij|u|ij) - (ij|u|ji)],$$

on substituting explicit spatial-orbital occupation numbers. This implies different effective Hamiltonians for closed and open shells, respectively, defined by functional derivatives

$$\frac{\delta(\Phi|H|\Phi)}{n_i^c \delta\chi_i^{c*}} = h\chi_i^c + \sum_j^c [2(j|u|j)\chi_i^c - (j|u|i)\chi_j^c]$$
$$+ \sum_j^o \left[(j|u|j)\chi_i^c - \frac{1}{2}(j|u|i)\chi_j^o\right] = \mathcal{H}^c\chi_i^c,$$

$$\frac{\delta(\Phi|H|\Phi)}{n_i^o \delta\chi_i^{o*}} = h\chi_i^o + \sum_j^c [2(j|u|j)\chi_i^o - (j|u|i)\chi_j^c]$$
$$+ \sum_j^o [(j|u|j)\chi_i^o - (j|u|i)\chi_j^o] = \mathcal{H}^o\chi_i^o.$$

For comparison, the UHF effective Hamiltonians, indexed by spin α and β, are

$$\frac{\delta(\Phi|H|\Phi)}{n_i^\alpha \delta\phi_i^{\alpha*}} = h\phi_i^\alpha + \sum_j^\alpha [(j|u|j)\phi_i^\alpha - (j|u|i)\phi_j^\alpha] + \sum_j^\beta (j|u|j)\phi_i^\alpha = \mathcal{H}^\alpha\phi_i^\alpha,$$

$$\frac{\delta(\Phi|H|\Phi)}{n_i^\beta \delta\phi_i^{\beta*}} = h\phi_i^\beta + \sum_j^\alpha (j|u|j)\phi_i^\beta + \sum_j^\beta [(j|u|j)\phi_i^\beta - (j|u|i)\phi_j^\beta] = \mathcal{H}^\beta\phi_i^\beta.$$

In a basis of RHF orbitals, $\mathcal{H}^c = \frac{1}{2}(\mathcal{H}^\alpha + \mathcal{H}^\beta)$ and $\mathcal{H}^o = \mathcal{H}^\alpha$. The constraint conditions $(\chi_i^c|\chi_j^c) = \delta_{ij}$, $(\chi_i^o|\chi_j^c) = 0$, $(\chi_i^o|\chi_j^o) = \delta_{ij}$ are incorporated into the RHF Euler–Lagrange equations for the spatial orbital functions

$$(\mathcal{H}^c - \epsilon_i^c)\chi_i^c = \sum_j^o \chi_j^o\lambda_{ji}^{oc}; \qquad (\mathcal{H}^o - \epsilon_i^o)\chi_i^o = \sum_j^c \chi_j^c\lambda_{ji}^{co},$$

using nonvanishing Lagrange multipliers $\lambda_{ji}^{oc} = (\chi_j^o|\mathcal{H}^c|\chi_i^c)$ and $\lambda_{ji}^{co} = (\chi_j^c|\mathcal{H}^o|\chi_i^o)$.

For the $1s^2 2s\ ^2S$ ground state of atomic Li, using the common notation $J_{n\ell}\chi_x = (n\ell|u|n\ell)\chi_x$, $K_{n\ell}\chi_x = (n\ell|u|x)\chi_{n\ell}$, the RHF Hamiltonians are

$$\mathcal{H}^c = h + 2J_{1s} + J_{2s} - K_{1s} - \frac{1}{2}K_{2s},$$
$$\mathcal{H}^o = h + 2J_{1s} + J_{2s} - K_{1s} - K_{2s},$$

and the UHF Hamiltonians are

$$\mathcal{H}^\alpha = h + 2J_{1s} + J_{2s} - K_{1s} - K_{2s},$$
$$\mathcal{H}^\beta = h + 2J_{1s} + J_{2s} - K_{1s}.$$

5.3.4 Algebraic Hartree–Fock: finite basis expansions

The OEL equations derived here as integrodifferential equations can be converted to linear algebraic equations by expanding the orbital functions in a basis set $\{\eta_p\}$ that can be extended to completeness in the orbital Hilbert space [347, 262, 349]. This expansion takes the general form $\phi_i = \sum_p \eta_p u_{pi}$, so that orbital variations $\delta\phi_i = \sum_p \eta_p\, \delta u_{pi}$ are specified by variations of the coefficients u_{pi}. If the basis set were complete, the variational condition $\int \delta\phi_i^* \{\mathcal{H} - \epsilon_i\}\phi_i\, d^3\mathbf{r} = 0$ would be equivalent to algebraic equations, $\sum_{p,q} \delta u_{pi}^* \int \eta_p^* \{\mathcal{H} - \epsilon_i\}\eta_q\, d^3\mathbf{r}\, u_{qi} = 0$. Free variation of the coefficients in a finite or countable basis implies matrix eigenvalue equations

$$\sum_q \{(p|\mathcal{H}|q) - \epsilon_i (p|q)\} u_{qi} = 0.$$

This is the stationary condition for the energy functional, when orbital functions are expanded in the specified basis set. RHF equations are related to UHF equations exactly as they are in the theory based on integrodifferential OEL equations.

5.3.5 Multiconfiguration SCF (MCSCF)

In some cases of strong correlation, or in general for open-shell states, an N-electron state may not be dominated by a single reference state. In these circumstances, an N-electron basis of Slater determinants Φ_μ constructed from an occupied set of more than N orbital functions defines a variational basis for the N-electron problem. A *configuration-interaction* (CI) calculation in the N-electron basis Φ_μ is combined with variational optimization of these occupied orbital functions. Following the general formalism developed here, a particular reference state $\Phi = \Phi_0$ is singled out as "first among equals", giving the expansion $\Psi = \Phi + \sum_{\mu\neq 0} \Phi_\mu c_\mu$ with the usual unsymmetrical normalization. The coefficients c_μ are determined by diagonalizing the N-electron matrix $H_{\mu\nu}$, or by constructing symmetry-adapted functions for open-shell configurations. Several of these coefficients may have unit magnitude. The energy functional for orbital variations is $E = (\Phi|H|\Phi) + \sum_{\mu\neq 0}(\Phi|H|\Phi_\mu)c_\mu$.

For this limited CI problem, it is convenient to diagonalize the one-electron density matrix, defining a set of *natural* orbitals that vary at each step of an iterative procedure. The one-electron density operator is defined by

$$\rho(\mathbf{r}, \mathbf{r}') = \frac{1}{(\Psi|\Psi)} \int \cdots \int d^3\mathbf{r}_2 \cdots d^3\mathbf{r}_N \Pi_{i=2}^N \delta(\mathbf{r}_i, \mathbf{r}'_i)$$
$$\times \Psi(\mathbf{r}_1, \ldots, \mathbf{r}_N)\Psi^*(\mathbf{r}'_1, \ldots, \mathbf{r}'_N)|_{\mathbf{r}=\mathbf{r}_1;\mathbf{r}'=\mathbf{r}'_1},$$

such that

$$\rho(\mathbf{r}, \mathbf{r}') = \sum_i \phi_i(\mathbf{r})\, d_{ij} \phi_j^*(\mathbf{r}').$$

Natural orbitals $\{\phi_i\}$ and occupation numbers n_i are determined by diagonalizing the density matrix d_{ij}. The reference state Φ is constructed from N orbitals with the largest occupation numbers. The energy functional is of the form

$$E = \sum_i n_i(i|h|i) + \frac{1}{2}\sum_{ijkl} \pi_{ijkl}(ij|\bar{u}|kl).$$

Orbital Euler–Lagrange equations are determined by functional derivatives

$$\frac{\delta E}{n_i \delta\phi_i^*(\mathbf{r})} = h\phi_i(\mathbf{r}) + \sum_{jkl} \frac{\pi_{ijkl}}{n_i}(j|\bar{u}|l)\phi_k(\mathbf{r}) = \mathcal{H}_i\phi_i(\mathbf{r}),$$

giving in general a different effective Hamiltonian for each orbital. The Euler–Lagrange equations are to solved subject to the orthonormality constraints $(i|j) = \delta_{ij}$. Introducing Lagrange multipliers $\lambda_{ji} = (j|\mathcal{H}_i|i) \neq \lambda_{ij} = (i|\mathcal{H}_j|j)$, these equations take the general form

$$\mathcal{H}_i\phi_i = \sum_j \phi_j\lambda_{ji}.$$

Orbital construction and diagonalization of the CI matrix are alternated until the calculations converge.

5.4 The optimized effective potential (OEP)

The computational effort of solving orbital Euler–Lagrange (OEL) equations is significantly reduced if the generally nonlocal exchange-correlation potential $\hat{v}_{xc}$ can be replaced or approximated by a local potential $v_{xc}(\mathbf{r})$. A variationally defined optimal local potential is determined using the optimized effective potential (OEP) method [380, 398]. This method can be applied to any theory in which the model exchange-correlation energy E_{xc} is an explicit orbital functional, so that the functional derivative $\delta E_{xc}/n_i\delta\phi_i^* = \hat{v}_{xc}\phi_i$ is specified for occupied orbital functions ϕ_i of a reference state Φ. Variations of the orbitals ϕ_i induce

$$\delta E = \sum_i n_i \int d^3\mathbf{r}\{\delta\phi_i^*(\mathbf{r})\mathcal{G}\phi_i(\mathbf{r}) + cc\},$$

expressed in terms of an effective Hamiltonian

$$\mathcal{G} = -\frac{1}{2}\nabla^2 + v(\mathbf{r}) + v_h(\mathbf{r}) + \hat{v}_{xc}$$

defined for the OEL equations by the functional derivative

$$\frac{\delta E}{n_i\delta\phi_i^*} = \mathcal{G}\phi_i.$$

A modified effective Hamiltonian $\mathcal{G}_{OEP}$ is defined by replacing $\hat{v}_{xc}$ by a model local potential $v_{xc}(\mathbf{r})$. The energy functional is made stationary with respect to variations of occupied orbitals ϕ_i that are determined by modified OEL equations in which $\mathcal{G}$ is replaced by $\mathcal{G}_{OEP}$. $\delta\phi_i$ is determined by variations $\delta v_{xc}(\mathbf{r})$ in these modified OEL equations. To maintain orthonormality, $\delta\phi_i$ can be constrained to be orthogonal to all occupied orbitals of the OEP trial state Φ, so that $\delta\phi_i(\mathbf{r}) = \sum_a(1-n_a)\phi_a(\mathbf{r})(a|\delta\phi_i)$. First-order perturbation theory for the OEP Euler–Lagrange equations implies that

$$\{\mathcal{G}_{OEP} - \epsilon_i\}\delta\phi_i = -\{\delta v_{xc} - (i|\delta v_{xc}|i)\}\phi_i,$$

or

$$(\epsilon_a - \epsilon_i)(a|\delta\phi_i) = -(a|\delta v_{xc}|i),$$

for $i \le N < a$. Expressed in terms of a Green's function, this implies

$$\delta\phi_i(\mathbf{r}) = -\int g_i(\mathbf{r},\mathbf{r}')\delta v_{xc}(\mathbf{r}')\phi_i(\mathbf{r}')\,d^3\mathbf{r}',$$

where

$$g_i(\mathbf{r},\mathbf{r}') = \sum_a (1-n_a)\phi_a(\mathbf{r})(\epsilon_a - \epsilon_i)^{-1}\phi_a^*(\mathbf{r}').$$

This defines $\delta\phi_i(\mathbf{r})/\delta v_{xc}(\mathbf{r}') = -g_i(\mathbf{r},\mathbf{r}')\phi_i(\mathbf{r}')$ as the kernel of a linear operator.

The variational equation induced by $\delta v_{xc} \to \delta\phi_i$ is

$$\sum_i n_i \sum_a (1-n_a)(i|\delta v_{xc}|a)(\epsilon_a - \epsilon_i)^{-1}(a|\mathcal{G}|i) = 0.$$

If δv_{xc} is arbitrary, this implies that

$$\sum_i n_i \phi_i^*(\mathbf{r}) \sum_a (1 - n_a)\phi_a(\mathbf{r})(\epsilon_a - \epsilon_i)^{-1}(a|\mathcal{G}|i) = 0.$$

On substituting $(a|\mathcal{G}_{OEP}|i) = 0$ from the OEP orbital equation, and using $(a|\mathcal{G}_{OEP} - \mathcal{G}|i) = (a|v_{xc} - \hat{v}_{xc}|i)$, this gives the OEP integral equation

$$\sum_i n_i \phi_i^*(\mathbf{r}) \sum_a (1 - n_a)\phi_a(\mathbf{r})(\epsilon_a - \epsilon_i)^{-1}(a|v_{xc} - \hat{v}_{xc}|i) = 0.$$

5.4.1 Variational formulation of OEP

The OEP integral equation is implied if the nonnegative integral

$$I = \frac{1}{2}\sum_i n_i \sum_a (1 - n_a)(i|v_{xc}(\mathbf{r}) - \hat{v}_{xc}|a)(\epsilon_a - \epsilon_i)^{-1}(a|v_{xc}(\mathbf{r}) - \hat{v}_{xc}|i),$$

evaluated for solutions of the OEP orbital equations, is stationary for variations of v_{xc}. If this local potential function is expanded in a basis set π_q, $v_{xc}(\mathbf{r}) = \sum_q \pi_q(\mathbf{r})\mathcal{L}_q$, the coefficient vector $\mathcal{L}_q$ is determined by linear algebraic equations [69, 151]

$$\sum_q \sum_{i,a} (i|p|a)(\epsilon_a - \epsilon_i)^{-1}(a|q|i)\mathcal{L}_q = \sum_{i,a} (i|p|a)(\epsilon_a - \epsilon_i)^{-1}(a|\hat{v}_{xc}|i),$$

where $(a|q|i) = \int \phi_a^*(\mathbf{r})\pi_q(\mathbf{r})\phi_i(\mathbf{r})d^3\mathbf{r}$. The variational formalism [151, 181] provides a practical methodology for molecular wave functions by introducing a basis-set expansion of the optimized local potential. Because only off-diagonal elements $(a|v_{xc}|i)$ occur in the OEP equations, the local effective potential is determined only up to an additive constant. The physical boundary condition that potentials for an isolated system must vanish at infinite distance fixes a specific value of this constant.

The variational energy $E_{OEP} = (\Phi_{OEP}|H|\Phi_{OEP})$ is minimized by the OEP equations subject only to the constraint of locality. Hence $E_{OEP} \geq E_{UHF}$, the Hartree–Fock ground-state energy. If an exact local exchange potential existed for UHF ground states, it would be determined by the OEP, implying $E_{OEP} = E_{UHF}$. Calculations for the closed-shell atoms He, Be, and Ne [1, 97] obtain OEP energies E_{OEP} and UHF energies E_{UHF} that agree for He, but E_{OEP}, in Hartree units, is -14.5724 for Be and -128.5455 for Ne, above E_{UHF}, -14.5730 for Be and -128.5471 for Ne, by amounts greater than the residual computational errors. When $E_{OEP} > E_{UHF}$, an exact local exchange potential cannot exist.

5.5 Density functional theory (DFT)

The density functional theory of Hohenberg, Kohn and Sham [173, 205] has become the standard formalism for first-principles calculations of the electronic structure of extended systems. Kohn and Sham postulate a model state described by a single-determinant wave function whose electronic density function is identical to the ground-state density of an interacting N-electron system. DFT theory is based on Hohenberg–Kohn theorems, which show that the external potential function $v(\mathbf{r})$ of an N-electron system is determined by its ground-state electron density. The theory can be extended to nonzero temperatures by considering a statistical electron density defined by Fermi–Dirac occupation numbers [241]. The theory is also easily extended to the spin-indexed density characteristic of UHF theory and of the two-fluid model of spin-polarized metals [414].

5.5.1 *The Hohenberg–Kohn theorems*

For N-electron ground states, Hohenberg and Kohn [173] (HK) proved theorems that remain valid for spin-indexed electron density functions $\rho(\mathbf{r})$ and external potential functions $v(\mathbf{r})$: (i) the electron density determines the external potential and hence the ground-state wave function Ψ and all physical properties; and (ii) a universal functional $F[\rho]$ is defined such that the energy functional $E_v = F + \int v\rho\, d^3\mathbf{r}$ is minimized by the ground-state density function, and yields the ground-state energy as its minimum value. The most straightforward proof, due to Levy [222], is a constrained-search construction. Given the N-electron Hamiltonian operator $H = \hat{T} + \hat{U} + \hat{V}$, where $\hat{V} = \sum_{i=1}^{N} v(\mathbf{r}_i)$, and a specified density function ρ normalized to N electrons, $\int \rho\, d^3\mathbf{r} = N$, and using the notation $\langle \cdots \rangle_t = (\Psi_t | \cdots | \Psi_t)/(\Psi_t|\Psi_t)$, then

$$F[\rho] = \min_{\Psi_t \to \rho} \langle \hat{T} + \hat{U} \rangle_t$$

defines a universal functional. Thus the mean value of $H - \hat{V} = \hat{T} + \hat{U}$ is minimized over all N-electron wave functions whose density function is the specified ρ. This requires that the given density must be physically realizable in a ground state. The constraint $\rho_t = \rho$ can be imposed using a Lagrange-multiplier field that is just the external potential function $v(\mathbf{r})$. Any trial external potential defines a functional

$$F_v[\rho] = \min_{\Psi_t} \left[\langle \hat{T} + \hat{U} \rangle_t + \int v(\rho_t - \rho) d^3\mathbf{r} \right] = E[v] - \int v\rho\, d^3\mathbf{r},$$

where $E[v]$ is the ground-state energy for external potential $v(\mathbf{r})$. The minimizing wave function is Ψ_v and the density is ρ_v. If $v = v_\rho$ is chosen such that $\rho_v = \rho$, this defines

$$F[\rho] = F_v[\rho_v] = E[v] - \int v\rho_v \, d^3\mathbf{r},$$

which can be computed in the ground state for any v.

A density function ρ is said to be "v-representable" if the Lagrange multiplier field $v = v_\rho$ exists. If so, this proves HK Theorem (i) by construction. v_ρ contains an arbitrary additive constant because $\int(\rho_t - \rho)d^3\mathbf{r} = N - N = 0$. This constant is fixed by the physical requirement that interactions vanish at infinite separation. For a nondegenerate ground state, v determines a unique density ρ. Uniqueness of the inverse mapping, from ρ to v, is proved by contradiction [173]. In the Levy construction [222], suppose that v_1 and v_2 define different ground-state wave functions Ψ_1 and Ψ_2 and distinct functionals F_1 and F_2 for the same ρ. The construction implies that $F_1[\rho] = E[v_1] - \int v_1\rho \leq F_2[\rho]$ because F_2 is just the mean value $\langle \hat{T} + \hat{U} \rangle_2$ evaluated for Ψ_2. But the theorem also implies $F_2[\rho] = E[v_2] - \int v_2\rho \leq F_1[\rho]$. This is a contradiction unless $v[\rho]$ and the functional $F[\rho]$ are uniquely defined for nondegenerate ground states. The Hohenberg–Kohn energy functional is $E_v[\rho] = F[\rho] + \int v\rho \, d^3\mathbf{r}$. Suppose that $F[\rho]$ corresponds to Ψ_ρ but $\rho \neq \rho_v$. Then

$$E_v[\rho] = \langle \hat{T} + \hat{U} + \hat{V} \rangle_\rho \geq E[v],$$

which proves HK Theorem (ii).

Since the exact ground-state electronic wave function and density can only be approximated for most N-electron systems, a variational theory is needed for the practical case exemplified by an orbital functional theory. As shown in Section 5.1, any rule $\Psi \to \Phi$ defines an orbital functional theory that in principle is exact for ground states. The reference state Φ for any N-electron wave function Ψ determines an orbital energy functional $E = E_0 + E_c$, in which $E_0 = T + E_h + E_x + V$ is a sum of explicit orbital functionals, and E_c is a residual correlation energy functional. In practice, the combination of exchange and correlation energy is approximated by an orbital functional E_{xc}.

The Levy construction [222] can be used to prove Hohenberg–Kohn theorems for the ground state of any such theory. It should be noted that any explicit model of the Hohenberg–Kohn functional $F[\rho]$ implies a corresponding orbital functional theory. The relevant density function $\rho(\mathbf{r})$ is that constructed from an OFT ground state. This has the orbital decomposition $\sum_i n_i\phi_i^*\phi_i$, as postulated by Kohn and Sham [205]. Unlike the density ρ_Ψ for an exact N-electron wave function Ψ, which cannot be determined for most systems of interest, the OFT ground-state density function is constructed from explicit solutions of the orbital Euler–Lagrange equations, and the theory is self-contained.

With fixed occupation numbers $n_i = 1$ for $i \leq N$, an orbital functional is a functional of a model or reference state Φ. For any model functional $E_{xc}[\Phi]$, an orbital functional $F = T + E_h + E_{xc}$ is defined such that $\min(F + V) = E[v]$ approximates an N-electron ground-state energy. The occupied orbital functions of the reference state Φ satisfy orbital Euler–Lagrange (OEL) equations of the form given above. Considering density functions ρ_Φ constructed from such orbital functions, suppose that one such function ρ is specified. The orbital functional $F[\Phi]$ is to be minimized subject to $\rho_\Phi = \rho$, following the Levy construction [222]. The density constraint is enforced using a Lagrange-multiplier field $v(\mathbf{r})$. For any given field $v(\mathbf{r})$ a functional of ρ is defined by

$$F_v[\rho] = \min_{\Phi_t}\left\{F[\Phi_t] + \int v(\rho_t - \rho)d^3\mathbf{r}\right\} = E[v] - \int v\rho\, d^3\mathbf{r}.$$

The minimizing model function Φ_v determines ρ_v. A ground-state density functional $F_s[\rho]$ is defined by $F_v[\rho]$ when v is chosen such that $\rho_v = \rho$. The notation F_s distinguishes a density functional restricted to ground states from the equivalent orbital functional F, defined for all functions in the orbital Hilbert space. This construction determines v for any given ρ if the variational problem has a solution. $E_v[\rho] = F_s[\rho] + \int v\rho\, d^3\mathbf{r}$ defines an energy functional for arbitrary v. When $\rho = \rho_v$, $E_v[\rho_v] = E[v]$, its minimum value. When $\rho \neq \rho_v$, $E_v[\rho] = F[\Phi] + \int v\rho\, d^3\mathbf{r}$ for some $\Phi(\to \rho) \neq \Phi_v$. Hence $E_v[\rho] \geq E_v[\rho_v]$. This establishes the variational property of $E_v[\rho]$, verifying the Hohenberg–Kohn theorems. Uniqueness of the Lagrange multiplier field v is implied for nondegenerate OFT solutions that minimize the orbital functional E.

5.5.2 Kohn–Sham equations

In any practical application of the Hohenberg–Kohn theory, a specified density functional $F_s[\rho]$ restricted to ground-state densities defines an equivalent orbital functional $F[\{\phi_i, n_i\}]$ that can be extended to all functions in the orbital Hilbert space. The OEL equations for occupied orbitals of the reference state of an N-electron ground state take the form, for $i \leq N$,

$$\{\hat{t} + v_h(\mathbf{r}) + \hat{v}_{xc}\}\phi_i = \{\epsilon_i - v(\mathbf{r})\}\phi_i,$$

derived from the orbital functional derivative

$$\frac{\delta F}{n_i \delta\phi_i^*} = \hat{f}\phi_i = \{\hat{t} + v_h(\mathbf{r}) + \hat{v}_{xc}\}\phi_i.$$

For exchange-correlation energy expressed as an orbital functional,

$$\hat{v}_{xc}\phi_i = \frac{\delta E_{xc}}{n_i \delta\phi_i^*}$$

defines a generally nonlocal potential $\hat{v}_{xc}$. The Kohn–Sham equations [205] for the model state corresponding to the same ground state are

$$\{\hat{t} + v_h(\mathbf{r}) + v_{xc}(\mathbf{r})\}\phi_i = \{\epsilon_i - v(\mathbf{r})\}\phi_i,$$

if a functional derivative $\frac{\delta E_{xc}}{\delta\rho} = v_{xc}(\mathbf{r})$ exists in the form of a local potential function. Assuming the existence of such a *Fréchet* functional derivative [26, 102] constitutes the *locality hypothesis*. If this hypothesis were valid, the OEL and Kohn–Sham equations would be equivalent, determining the same model or reference state.

It might appear paradoxical to invoke the locality hypothesis for $v_{xc}(\mathbf{r})$ while using the Schrödinger operator $\hat{t}$ rather than assumed Fréchet derivative $\frac{\delta T}{\delta\rho} = v_T(\mathbf{r})$ for the kinetic energy. Applying Hohenberg–Kohn theory to a noninteracting N-electron system, Kohn and Sham [205] show that the kinetic energy can be expressed as a functional $T_s[\rho]$ of the same ground-state density used to define $F_s[\rho]$. Following the constrained-search logic of Levy [222], the density constraint can be enforced by a Lagrange multiplier field $w(\mathbf{r})$ that acts as a local effective potential in the Kohn–Sham orbital equations. Thus it would appear that ground-state theory for an interacting N-electron system can be replaced by a noninteracting model constructed to obtain the same density, and that the theory requires only a local potential $w(\mathbf{r})$. Such a conclusion depends on the locality hypothesis, that Fréchet functional derivatives exist for density functionals. This question is most easily examined for a noninteracting system in the context of Thomas–Fermi theory, where only the kinetic energy is relevant. Before examining this issue, consistency conditions are derived for orbital and density functional derivatives of the same functional.

5.5.3 Functional derivatives and local potentials

The variation of an orbital functional induced by infinitesimal orbital variations is

$$\delta F = \int d^3\mathbf{r} \sum_i n_i \left\{ \delta\phi_i^*(\mathbf{r}) \frac{\delta F}{n_i \delta\phi_i^*(\mathbf{r})} + cc \right\}.$$

If this becomes a density functional $F_s[\rho]$ for ground states, the density functional variation is

$$\delta F_s = \int d^3\mathbf{r} \sum_i n_i \left\{ \delta\phi_i^*(\mathbf{r}) \frac{\delta F_s}{\delta\rho(\mathbf{r})} \phi_i(\mathbf{r}) + cc \right\}.$$

Consistency between these variations of a density functional and of the equivalent orbital functional implies the chain rule $\frac{\delta F_s}{\delta\rho(\mathbf{r})}\phi_i(\mathbf{r}) = \frac{\delta F}{n_i\delta\phi_i^*(\mathbf{r})}$, regardless of any constraint on variations. This implies the sum rule

$$\sum_i n_i\phi_i^* \frac{\delta F_s}{\delta\rho}\phi_i = \sum_i n_i\phi_i^* \frac{\delta F}{n_i\delta\phi_i^*}.$$

An orbital functional derivative in general defines a linear operator such that $\frac{\delta F}{n_i\delta\phi_i^*} = \hat{v}_F\phi_i$. If the locality hypothesis were valid, then $\frac{\delta F_s}{\delta\rho(\mathbf{r})} = v_F(\mathbf{r})$, and the implied local potential function could be computed directly from the sum rule,

$$v_F(\mathbf{r})\rho(\mathbf{r}) = \sum_i n_i\phi_i^*\hat{v}_F\phi_i.$$

This formula was used by Slater [385] to define an effective local exchange potential. The generally unsatisfactory results obtained in calculations with this potential indicate that the locality hypothesis fails for the density functional derivative of the exchange energy E_x [294].

5.5.4 Thomas–Fermi theory

The Hohenberg–Kohn theory of N-electron ground states is based on consideration of the spin-indexed density function. Much earlier in the development of quantum mechanics, Thomas–Fermi theory [402, 108] (TFT) was formulated as exactly such a density-dependent formalism, justified as a semiclassical statistical theory [231, 232]. Since Hohenberg–Kohn theory establishes the existence of an exact universal functional $F_s[\rho]$ for ground states, it apparently implies the existence of an exact ground-state Thomas–Fermi theory. The variational theory that might support such a conclusion is considered here.

The orbital theory of Kohn and Sham [205] differs from Thomas–Fermi theory in that the density function $\rho = \sum_i n_i\rho_i = \sum_i n_i\phi_i^*\phi_i$ is postulated to have an orbital structure, with occupation numbers n_i that are consistent with Fermi–Dirac statistics. The orbital functions ϕ_i are determined by orbital Kohn–Sham (KS) equations, in which the kinetic energy is represented by the operator $\hat{t} = -\frac{1}{2}\nabla^2$ of Schrödinger. The Thomas–Fermi equations ignore such orbital structure and represent kinetic energy by a local effective potential $v_T(\mathbf{r})$, defined in principle as the functional derivative $\delta T_s/\delta\rho$ of a kinetic-energy functional $T_s[\rho]$. If such a Fréchet derivative exists, the two theories are just alternative ways of describing the same ground state.

In Kohn–Sham theory, densities are postulated to be sums of orbital densities, for functions ϕ_i in the orbital Hilbert space. This generates a Banach space [102] of density functions. Thomas–Fermi theory can be derived if an energy functional $E[\rho] = F_s[\rho] + V[\rho]$ is postulated to exist, defined for all normalized ground-state densities in this density function space, including all infinitesimal neighborhoods of such densities. Introducing a Lagrange multiplier μ to enforce the normalization constraint $\int \rho\, d^3\mathbf{r} = N$, the variational equation is

$$\delta\{F_s[\rho] + V[\rho] - \mu(\int \rho\, d^3\mathbf{r} - N)\}] = 0,$$

or, if a Fréchet functional derivative $\frac{\delta F_s}{\delta \rho}$ exists,

$$\int \delta\rho \left\{ \frac{\delta F_s}{\delta \rho} + v(\mathbf{r}) - \mu \right\} d^3\mathbf{r} = 0.$$

If F_s is defined for unrestricted variations of ρ in any infinitesimal function neighborhood of a solution, this implies the Thomas–Fermi (TF) equation

$$\frac{\delta F_s}{\delta \rho} = \mu - v(\mathbf{r}).$$

The Lagrange multiplier μ, determined by normalization, is the chemical potential [232], such that $\mu = \partial E/\partial N$ when the indicated derivative is defined. This derivation requires the locality hypothesis, that a Fréchet derivative of $F_s[\rho]$ exists as a local function $v_F(\mathbf{r})$.

The locality hypothesis can be tested in a noninteracting model, in which the functional F_s is replaced by T_s. The kinetic energy orbital functional is $T = \sum_i n_i(i|\hat{t}|i)$ and the OEL equations are just orbital Schrödinger equations

$$\hat{t}\phi_i = \{\epsilon_i - v(\mathbf{r})\}\phi_i. \tag{5.8}$$

If the locality hypothesis is valid, then $\frac{\partial T_s}{\partial \rho} = v_T(\mathbf{r})$, and the Thomas–Fermi equation is

$$v_T(\mathbf{r}) = \mu - v(\mathbf{r}). \tag{5.9}$$

The trace sum $\sum_i n_i(i|v_T - \hat{t}|i)$ must vanish if the OEL and TF equations are equivalent. Equations (5.8) and (5.9), multiplied by appropriate factors, summed over orbitals, and integrated, imply the sum rule [288]

$$\sum_i n_i\epsilon_i = N\mu.$$

Since $\sum_i n_i = N$ and all $\epsilon_i \leq \mu$, this implies that all ϵ_i are equal, in violation of the exclusion principle for any compact system with more than two electrons. The failure of this sum rule implies that in general the assumed Fréchet derivative of $T_s[\rho]$ cannot exist for more than two electrons, and there can be no exact Thomas–Fermi theory.

If the hypothetical Fréchet derivative $v_T(\mathbf{r})$ could be replaced by the operator $\hat{t}$ when acting on occupied orbital functions ϕ_i, there would be no contradiction. It can be shown that this is the correct implication of variational theory. The functional derivative in question can be shown to be a *Gâteaux* derivative [26, 102], the analog in functional analysis of a partial derivative, equivalent to a linear operator that acts on orbital functions. Consider variations of T_s induced by unconstrained orbital variations about ground-state solutions of the OEL equations

$$\delta T_s = \int d^3\mathbf{r} \sum_i n_i(\delta\phi_i^* \hat{t}\phi_i + cc) = \int d^3\mathbf{r} \sum_i n_i(\delta\phi_i^* \{\epsilon_i - v\}\phi_i + cc)$$
$$= \int d^3\mathbf{r} \sum_i n_i \{\epsilon_i - v(\mathbf{r})\}\delta\rho_i(\mathbf{r}).$$

Since $\rho = \sum_i n_i \rho_i$, this equation determines partial functional derivatives

$$\frac{\delta T_s}{n_i \delta\rho_i} = \epsilon_i - v(\mathbf{r}). \tag{5.10}$$

This defines a Gâteaux functional derivative [26, 102], whose value depends on a "direction" in the function space, reducing to a Fréchet derivative only if all ϵ_i are equal. Defining $\mathcal{H} = \hat{t} + v$, an explicit orbital index is not needed if Eq. (5.10) is interpreted to define a linear operator acting on orbital wave functions, $\mathcal{H} - v = \hat{t}$. The elementary chain rule $\frac{\delta T_s}{n_i \delta\rho_i} = \frac{\partial\rho}{n_i \partial\rho_i}\frac{\delta T_s}{\delta\rho} = \frac{\delta T_s}{\delta\rho}$ is valid when the functional derivatives are interpreted as linear operators. This confirms the chain rule, $\frac{\delta T_s}{\delta\rho}\phi_i = \frac{\delta T}{n_i \delta\phi_i^*} = \hat{t}\phi_i$.

This argument shows that the locality hypothesis fails for more than two electrons because the assumed Fréchet derivative must be generalized to a Gâteaux derivative, equivalent in the context of OEL equations to a linear operator that acts on orbital wave functions. The conclusion is that the use by Kohn and Sham of Schrödinger's operator $\hat{t}$ is variationally correct, but no equivalent Thomas–Fermi theory exists for more than two electrons. Empirical evidence (atomic shell structure, chemical binding) supports the Kohn–Sham choice of the nonlocal kinetic energy operator, in comparison with Thomas–Fermi theory [288]. A further implication is that if an explicit approximate local density functional E_{xc} is postulated, as in the local-density approximation (LDA) [205], the resulting Kohn–Sham theory is variationally correct. Typically, for $E_{xc} = \int e_{xc}(\rho)\rho\, d^3\mathbf{r}$, the density functional derivative is a Fréchet derivative, the local potential function $v_{xc} = e_{xc} + \rho\, de_{xc}/d\rho$.

5.5.5 The Kohn–Sham construction

A particular mapping $\Psi \to \Phi$ is determined by the Kohn–Sham construction (KSC): minimize the kinetic energy orbital functional $T = \sum_i (i|\hat{t}|i)$ for specified spin-indexed electron density ρ. This applies Hohenberg–Kohn logic to a noninteracting system and constructs a ground-state kinetic energy functional $T_s[\rho]$. If the Levy constrained-search algorithm has a solution, the specified density ρ is said to be "noninteracting v-representable", and a Lagrange-multiplier field $w(\mathbf{r})$ is determined that acts as a local potential in the resulting noninteracting Kohn–Sham equations. If the specified density is that of a nondegenerate ground state of an interacting system, and if $E_{xc}[\rho]$ for this system has a Fréchet functional derivative that defines a local potential $v_{xc}(\mathbf{r})$, then Hohenberg–Kohn logic implies that the orbital equations for the interacting and noninteracting systems must be equivalent. Because the local potentials must be equal if both vanish at infinity, $v_{xc}(\mathbf{r})$ is determined by subtraction:

$$v_{xc}(\mathbf{r}) = w(\mathbf{r}) - v(\mathbf{r}) - v_h(\mathbf{r}).$$

Hohenberg–Kohn theorems for any orbital functional model imply that the Kohn–Sham construction must result in noninteracting KS equations equivalent to the OEL equations if the latter contain only local potential functions. The relevant density function is that computed from the orbital functional ground state. Since the exchange-correlation term in the OEL equations is $\frac{\delta E_{xc}}{n_i \delta \phi_i^*} = \hat{v}_{xc}\phi_i$, the existence of a Fréchet derivative $v_{xc}(\mathbf{r})$ is not assured, and must be proven for each particular model of E_{xc}. However, $v_{xc}(\mathbf{r})$ always exists in the local-density approximation. If a local potential v_{xc} does not exist, there is no clear relationship between the noninteracting orbital equations obtained by the KS construction and the ground-state OEL equations which minimize the energy functional.

The unrestricted Hartree–Fock theory (UHF) for closed-shell atoms provides an exchange-only orbital model for which Hohenberg–Kohn theorems can be proved [324]. Ground-state orbital wave functions and energies and total variational energy are known to high numerical accuracy [130]. The Kohn–Sham construction (KSC) has been carried out for the atoms He, Be, and Ne with sufficient numerical accuracy to test consequences of the locality hypothesis for the exchange potential [152, 294]. These calculations verify by construction that the UHF densities have the property of "noninteracting v-representability", but the computed wave functions are not the ground states of the variational UHF model, for more than two electrons. Results can be compared with OEP calculations (optimized effective potential) for these atoms [1, 97]. Because both KSC and OEP constrain the exchange potential to be local, while KSC also constrains the density function, these nested variational conditions imply that $E_{KSC} \geq E_{OEP} \geq E_{UHF}$ [152]. The computed energies are shown in Table (5.1). These computed energies are consistent with the existence of a local exchange potential for He but not for Be and Ne. This confirms the discussion given above, which indicates that Fréchet functional derivatives can exist for two electrons, but not in general otherwise, because density functional derivatives differ for partial orbital densities that correspond to different orbital energies.

Table 5.1. *Computed total energies for typical atoms (Hartree units)*

Atom	E_{UHF}	E_{OEP}	E_{KSC}
He	−2.8617	−2.8617	−2.8617
Be	−14.5730	−14.5724	−14.5724
Ne	−128.5471	−128.5455	−128.5454

If an exact local exchange potential does not exist, there is no reason for UHF, OEP, and KSC results to be the same in the UHF model. That the OEP density is not exactly equal to that of the UHF ground state is indicated by an analysis of OEP results [412], and by recent test calculations [69]. This would imply that the density constraint in KSC is a true variational constraint for more than two electrons, so that $E_{KSC} > E_{OEP}$. The calculations considered here may not have sufficient numerical accuracy to establish this evidently small energy difference.

6

Time-dependent theory and linear response

The principal references for this chapter are:

[79] Dirac, P.A.M. (1930). Note on exchange phenomena in the Thomas atom, *Proc. Camb. Phil. Soc.* **26**, 376–385.

[94] Ehrenreich, H. and Cohen, M.H. (1959). Self-consistent field approach to the many-electron problem, *Phys. Rev.* **115**, 786–790.

[155] Gross, E.K.U., Dobson, J.F. and Petersilka, M. (1996). Density functional theory of time-dependent phenomena, in: *Density Functional Theory*, ed. R.F. Nalewajski (Springer, Berlin).

[238] McLachlan, A.D. and Ball, M.A. (1964). Time-dependent Hartree-Fock theory for molecules, *Rev. Mod. Phys.* **36**, 844–855.

[292] Nesbet, R.K. (2001). Orbital functional theory of linear response and excitation, *Int. J. Quantum Chem.* **86**, 342–346.

[329] Petersilka, M., Gossmann, U.J. and Gross, E.K.U. (1996). Excitation energies from time-dependent density-functional theory, *Phys. Rev. Lett.* **76**, 1212–1215.

[351] Runge, E. and Gross, E.K.U. (1984). Density-functional theory for time-dependent systems, *Phys. Rev. Lett.* **52**, 997–1000.

[407] Thouless, D.J. (1961). *The Quantum Mechanics of Many-body Systems* (Academic Press, New York).

In the quantum theory of interacting electrons, a physically correct theory of time dependence should in principle be formulated as a relativistic quantum field theory. The physical model is that of electrons, each characterized by a probability distribution over space-time events $\mathbf{x}_i, t_i$, that interact indirectly through the quantized electromagnetic field. This theory is simplified for particular applications by neglecting true radiative effects of quantum electrodynamics, and by passing to the limit of large c, the velocity of light *in vacuo*. Although this theory cannot be developed with adequate detail in the present context, the discussion will emphasize independent-electron models that are consistent with a physical picture of instantaneous direct interactions replaced by a mean field that varies with a time parameter that is the same for all electrons.

6.1 The time-dependent Schrödinger equation for one electron

For a single electron, the time-independent Schrödinger eigenvalue problem is determined by the variational condition

$$\int d^3\mathbf{x}[\delta\psi^*\{\mathcal{H}-\epsilon\}\psi + cc] = 0,$$

for variations of trial functions ψ that belong to the usual Hilbert space and satisfy specified boundary conditions on the surface of the volume of integration. The Lagrange multiplier ϵ is determined so that the normalization integral $\int d^3\mathbf{x}\psi^*\psi$ remains constant. The resulting Schrödinger equation

$$\{\mathcal{H}-\epsilon\}\psi = 0,$$

is modified in time-dependent theory to

$$i\hbar\frac{\partial}{\partial t}\psi(xt) = \mathcal{H}(xt)\psi(xt).$$

The simplified notation (xt) is used here to denote $(\mathbf{x}, t)$. If $\mathcal{H}$ is independent of time, these two equations are equivalent. The wave function $\psi(x)$ is modified by a time-dependent phase factor $\exp(\epsilon t/i\hbar)$, which has no physical consequences.

Following Hamilton's principle in classical mechanics, the required time dependence can be derived from a variational principle based on a seemingly artificial Lagrangian density, integrated over both space and time to define the functional

$$A[\psi] = \int_{t_0}^{t_1} dt \int d^3\mathbf{x}\psi^*\left\{i\hbar\frac{\partial}{\partial t} - \mathcal{H}\right\}\psi.$$

Treating variations of ψ and ψ^* as independent, because they lead to equivalent Euler–Lagrange equations, and integrating the time integral by parts as in Euler's theory, the resulting variational expression is

$$\delta A = \int_{t_0}^{t_1} dt \int d^3\mathbf{x}\left[\delta\psi^*\left\{i\hbar\frac{\partial}{\partial t} - \mathcal{H}\right\}\psi + cc\right] = 0,$$

if the normalization integral $\int d^3\mathbf{x}\psi^*\psi$ is held constant. This implies the time-dependent Schrödinger equation in the time interval $t_0 \leq t \leq t_1$.

The action integral A is not changed if the trial function ψ is multiplied by a phase factor $\exp(\int^t \gamma(t')dt'/i\hbar)$, while $\mathcal{H}$ is increased by a time-dependent but spatially uniform potential $\gamma(t)$. This is an example of *gauge invariance*, taken out of the usual context of electromagnetic theory. Indicating the modified wave function by ψ_γ, the modified action integral is

$$A_\gamma = \int_{t_0}^{t_1} dt \int d^3\mathbf{x}\left[\delta\psi_\gamma^*\left\{i\hbar\frac{\partial}{\partial t} - \mathcal{H} - \gamma(t)\right\}\psi_\gamma + cc\right].$$

An alternative interpretation of this equation is that $\gamma(t)$ is a time-dependent Lagrange multiplier, introduced to enforce constant normalization $(\psi|\psi)$. By the conventional definition of an isolated system, asymptotic values of potential functions in $\mathcal{H}$ should become negligible outside some enclosing surface. Since for finite systems this is a property of potentials derived from Coulombic interactions, the otherwise arbitrary function $\gamma(t)$ is set to zero for such potentials.

6.2 The independent-electron model as a quantum field theory

The reference state Φ of *N*-electron theory becomes a reference vacuum state $|\Phi\rangle$ in the field theory. A complete orthonormal set of spin-indexed orbital functions $\phi_p(\mathbf{x})$ is defined by eigenfunctions of a one-electron Hamiltonian $\mathcal{H}$, with eigenvalues ϵ_p. The reference vacuum state corresponds to the ground state of a noninteracting *N*-electron system determined by this Hamiltonian. *N* occupied orbital functions $(\epsilon_i \leq \mu)$ are characterized by fermion creation operators $a_i^\dagger$ such that $a_i^\dagger|\Phi\rangle = 0$. Here μ is the chemical potential or Fermi level. A complementary orthogonal set of unoccupied orbital functions are characterized by destruction operators a_a such that $a_a|\Phi\rangle = 0$ for $\epsilon_a > \mu$ and $a > N$. A fermion quantum field is defined in this orbital basis by

$$\psi(\mathbf{x}, t) = \sum_p \phi_p(\mathbf{x}) a_p(t).$$

The fermion creation and destruction operators are defined such that $a_p a_q^\dagger + a_q^\dagger a_p = \delta_{pq}$. In analogy to relativistic theory, and more appropriate to the linear response theory to be considered here, the elementary fermion operators a_p can be treated as algebraic objects fixed in time, while the orbital functions are solutions of a time-dependent Schrödinger equation

$$i\hbar\frac{\partial}{\partial t}\phi_p(xt) = \mathcal{H}(xt)\phi_p(xt).$$

The fermion field operator takes the form

$$\psi(xt) = \sum_p \phi_p(xt) a_p.$$

Occupation numbers are defined by $n_p = \langle\Phi|a_p^\dagger a_p|\Phi\rangle$, such that $n_i = 1$, $n_a = 0$ for $i \leq N < a$. An $(N-1)$-electron basis state is defined such that $|\Phi_i\rangle = a_i|\Phi\rangle$ if $i \leq N$. The orthonormality condition is verified by

$$\langle\Phi_i|\Phi_j\rangle = \langle\Phi|a_i^\dagger a_j|\Phi\rangle = n_i\delta_{ij}.$$

An $(N+1)$-electron basis state is defined such that $|\Phi^a\rangle = a_a^\dagger|\Phi\rangle$ if $a > N$. Here the orthonormality condition is

$$\langle \Phi^a | \Phi^b \rangle = \langle \Phi | a_a a_b^\dagger | \Phi \rangle = (1 - n_a)\delta_{ab}.$$

These definitions remain valid in relativistic theory.

A Dirac density operator is defined at specified time by its matrix kernel

$$\hat{\rho}(\mathbf{x}, \mathbf{x}') = \sum_p \phi_p(\mathbf{x}) n_p \phi_p^*(\mathbf{x}') = \langle \Phi | \psi^\dagger(\mathbf{x}') \psi(\mathbf{x}) | \Phi \rangle.$$

The Dirac density operator for the reference state is idempotent:

$$\hat{\rho}^2 = \sum_{p,q} \phi_p n_p (p|q) n_q \phi_q^* = \sum_p \phi_p(\mathbf{x}) n_p^2 \phi_p^*(\mathbf{x}') = \hat{\rho},$$

because $n_p^2 = n_p$ for either occupied or unoccupied orbitals. Normalization is such that $\mathrm{Tr}\hat{\rho} = \sum_p n_p = N$.

Reduction of the time-dependent N-electron problem to an independent-electron model appears to be consistent with quantum field theory, if true radiative effects are neglected. In this approximation, the theory can be developed in Coulomb gauge, with instantaneous interactions. Except for effects due to quasiparticle lifetimes, which must be considered whenever there are continuum states, the mean field of the independent-particle model changes instantaneously with changes of local electron density. The orbital wave functions satisfy Schrödinger or Dirac equations determined by a time-dependent effective mean field. It appears plausible that the time development of such a system should be described as in time-independent theory at each instant, carried forward in time by a continuously developing unitary change of representation of the self-consistent orbital basis. This is the physical implication of Dirac's density operator, which remains idempotent so long as the effective Hamiltonian is Hermitian.

Ultimately, the theory must be consistent with quantum electrodynamics, which reduces in the absence of radiative terms to a time-dependent equation, in which ψ is interpreted as a fermion field operator $\sum \phi_p a_p$ acting on the state function $|\Phi\rangle$. The field equation of motion is equivalent to simultaneous equations

$$\left\{ i\hbar \frac{\partial}{\partial t} - \mathcal{G} \right\} \phi_i(xt) a_i |\Phi\rangle = 0,$$

for noninteracting electrons in the self-consistent mean field described by an operator $\mathcal{G}$. In the nonrelativistic orbital theories considered here, $\mathcal{G}$ reduces to an effective one-electron Hamiltonian operator.

6.3 Time-dependent Hartree–Fock (TDHF) theory

6.3.1 Operator form of Hartree–Fock equations

When the idempotent density operator $\hat{\rho}$ is constructed from orbital solutions of the Hartree–Fock equations, $(\mathcal{H} - \epsilon_i)\phi_i = 0$, it satisfies the commutator equation

$$[\mathcal{H}, \hat{\rho}] = \mathcal{H}\hat{\rho} - \hat{\rho}\mathcal{H} = 0.$$

This follows from

$$\mathcal{H}\hat{\rho} = \sum_i \{\mathcal{H}\phi_i(\mathbf{x})\} n_i \phi_i^*(\mathbf{x}') = \sum_i \phi_i(\mathbf{x})\epsilon_i n_i \phi_i^*(\mathbf{x}').$$

Similarly, operating to the left with $\mathcal{H}$,

$$\hat{\rho}\mathcal{H} = \sum_i \phi_i(\mathbf{x}) n_i \epsilon_i^* \phi_i^*(\mathbf{x}').$$

Hence, for real eigenvalues, $\mathcal{H}\hat{\rho} - \hat{\rho}\mathcal{H} = 0$.

As shown by Dirac [79], the corresponding time-dependent equation takes the form

$$i\hbar\frac{\partial}{\partial t}\hat{\rho}(t) = \mathcal{H}(t)\hat{\rho}(t) - \hat{\rho}(t)\mathcal{H}(t) = [\mathcal{H}(t), \hat{\rho}(t)],$$

consistent with a time-dependent Hartree–Fock equation,

$$i\hbar\frac{\partial}{\partial t}\phi_i = \mathcal{H}\phi_i.$$

The time-dependent Hartree–Fock equation is expressed formally by

$$\phi_i(xt) = e^{\int_0^t \mathcal{H}(u)du/i\hbar}\phi_i(x0).$$

This implies consistency conditions for Hermitian $\mathcal{H}$,

$$\int_y \phi_i^*(yt)\phi_j(yt) = \int_y \left(e^{\int_0^t \mathcal{H}(u)du/i\hbar}\phi_i(y0)\right)^* \left(e^{\int_0^t \mathcal{H}(u)du/i\hbar}\phi_j(y0)\right)$$

$$= \int_y \phi_i^*(y0)e^{\int_0^t(\mathcal{H}(u)-\mathcal{H}^\dagger(u))du/i\hbar}\phi_j(y0) = \delta_{ij}.$$

If $\mathcal{H}(t) \equiv \mathcal{H}(0)$, $\phi_i(xt) = e^{\epsilon_i t/i\hbar}\phi_i(x0)$.

6.3.2 The screening response

For a stationary state at $t = 0$, $[\mathcal{H}(0), \hat{\rho}(0)] = 0$. When $t \geq 0$, a weak perturbing potential $\Delta v(t)$ modifies the effective Hamiltonian by an induced screening term such that $\Delta\mathcal{H} = \Delta v + \Delta\mathcal{H}_s$. The first-order perturbation equations for $\Delta\rho(t) = \hat{\rho}(t) - \hat{\rho}(0)$ are

$$i\hbar\frac{\partial}{\partial t}\Delta\rho = [\mathcal{H}(0), \Delta\rho] + [\Delta\mathcal{H}, \hat{\rho}(0)].$$

In a basis set of occupied and unoccupied orbitals of a reference state Φ, variation of an occupied orbital ϕ_i affects Φ only through an incremental sum of unoccupied orbitals $\Delta\phi_i(t) = \sum_a (1 - n_a)\phi_a c_i^a(t)$. The corresponding variation of the density matrix is

$$\Delta\rho(\mathbf{x}, \mathbf{x}'; t) = \sum_i \sum_a n_i(1 - n_a)\big[\phi_a(\mathbf{x})c_i^a(t)\phi_i^*(\mathbf{x}') + \phi_i(\mathbf{x})c_i^a(t)^*\phi_a^*(\mathbf{x}')\big].$$

From the Fock operator $\sum_j n_j(j|\bar{u}|j)$,

$$\Delta\mathcal{H}_s(t) = \sum_j \sum_b n_j(1 - n_b)\big[(j|\bar{u}|b)c_j^b(t) + (b|\bar{u}|j)c_j^b(t)^*\big].$$

Using $(p|\mathcal{H}(0)|j) = \epsilon_p\delta_{pj}$ and $(j|\hat{\rho}(0)|q) = n_q\delta_{jq}$,

$$(p|[\mathcal{H}(0), \Delta\rho]|q) = (\epsilon_p - \epsilon_q)(p|\Delta\rho|q);$$
$$(p|[\Delta\mathcal{H}, \hat{\rho}(0)]|q) = (n_q - n_p)(p|\Delta\mathcal{H}|q).$$

The linear response to $\Delta v(xt) = 2\Delta w(\mathbf{x})\Re(e^{-i\omega t})$, a weak time-dependent perturbing potential, is given for indices $i, j \leq N < a, b$ by

$$\begin{aligned} i\hbar\dot{c}_i^a(t) &= (\epsilon_a - \epsilon_i)c_i^a(t) + (n_i - n_a)(a|\Delta v(xt)|i) \\ &\quad + (n_i - n_a)\sum_j \sum_b \big[(aj|\bar{u}|ib)c_j^b(t) + (ab|\bar{u}|ij)c_j^b(t)^*\big]. \end{aligned}$$

For $c_i^a(t) = X_i^a e^{-i\omega t} + Y_i^{a*} e^{i\omega^* t}$, treating frequency ω as a complex number,

$$\begin{aligned} (\epsilon_a - \epsilon_i - \hbar\omega)X_i^a + (n_i - n_a)\sum_j \sum_b \big[(aj|\bar{u}|ib)X_j^b + (ab|\bar{u}|ij)Y_j^b\big] \\ = -(n_i - n_a)(a|\Delta w|i), \\ (\epsilon_a - \epsilon_i + \hbar\omega)Y_i^a + (n_i - n_a)\sum_j \sum_b \big[(ij|\bar{u}|ab)X_j^b + (ib|\bar{u}|aj)Y_j^b\big] \\ = -(n_i - n_a)(i|\Delta w|a). \end{aligned}$$

If there is no driving term $\Delta w(\mathbf{x})$ in these TDHF equations, discrete excitation energies are determined by values of $\hbar\omega$ for which the determinant of the residual homogeneous equations vanishes. This gives the equations of the random-phase approximation (RPA) for excitation energies, in an exact-exchange model [94, 407].

6.4 Time-dependent orbital functional theory (TOFT)

6.4.1 Remarks on time-dependent theory

Dirac's development of TDHF theory invokes the Heisenberg equation of motion for the density operator as a basic postulate,

$$i\hbar\frac{\partial}{\partial t}\hat{\rho} = [\mathcal{H}, \hat{\rho}].$$

Equivalently, time-dependent canonical Hartree–Fock equations are assumed to take the same form as the time-dependent Schrödinger equation.

These concepts, inherent in the TDHF formalism, generalize immediately to orbital functional theory, when electronic correlation energy is included in the model. Given some definition $\Psi \to \Phi$ that determines a reference state Φ for any N-electron state Ψ, correlation energy can be defined for any stationary state by $E_c = E - E_0$, where $E_0 = (\Phi|\hat{H}|\Phi)$ and $E = (\Phi|\hat{H}|\Psi)$. Conventional normalization $(\Phi|\Psi) = (\Phi|\Phi) = 1$ is assumed. A formally exact functional $E_c[\Phi]$ exists for stationary states, for which a mapping $\Phi \to \Psi$ is established by the Schrödinger equation [292]. Because both Φ and $\hat{\rho}$ are defined by the occupied orbital functions $\{\phi_i\}$, for fixed occupation numbers n_i, $E[\Phi]$, $E[\hat{\rho}]$ and $E[\{\phi_i\}]$ are equivalent functionals. Since E_0 is an explicit orbital functional, any approximation to E_c as an orbital functional defines a TOFT theory. Because a formally exact functional E_c exists for stationary states, linear response of such a state can also be described by a formally exact TOFT theory. In nonperturbative time-dependent theory, total energy is defined only as a mean value $E(t)$, which lies outside the range of definition of the exact orbital functional $E_c[\{\phi_i\}]$ for stationary states. Although this may preclude a formally exact TOFT theory, the formalism remains valid for any model based on an approximate functional E_c.

Any postulated orbital functional $E[\{\phi_i\}]$ defines an action integral,

$$A[\{\phi_i\}] = \int_{t_0}^{t_1} dt\left(\sum_i n_i \int d^3\mathbf{x}\,\phi_i^* i\hbar\frac{\partial\phi_i}{\partial t} - E[\{\phi_i\}]\right),$$

that is stationary for $t_0 \leq t \leq t_1$ if and only if the occupied orbital functions satisfy the time-dependent Euler–Lagrange equations

$$i\hbar\frac{\partial\phi_i(xt)}{\partial t} = \frac{\delta E}{n_i\delta\phi_i^*(xt)} = \mathcal{G}(xt)\phi_i(xt).$$

Here $\mathcal{G} = \mathcal{H} + \hat{v}_c$, where $\mathcal{H}$ is defined by $\delta E_0/n_i\delta\phi_i^* = \mathcal{H}\phi_i$ as in Hartree–Fock theory. $\hat{v}_c$ is defined by the functional derivative $\frac{\delta E_c}{n_i\delta\phi_i^*} - \hat{v}_c\phi_i$. If E_c is omitted, the theory reduces to TDHF.

6.4.2 Exact linear response theory

When extended to include electronic correlation, for which an exact but implicit orbital functional was derived above, the TDHF formalism becomes a formally exact theory of linear response. In practice, some simplified orbital functional $E_c[\{\phi_i\}]$ must be used, and the accuracy of results is limited by this choice. The Hartree–Fock operator $\mathcal{H}$ is replaced by $\mathcal{G} = \mathcal{H} + \hat{v}_c$. Dirac defines an idempotent density operator $\hat{\rho}$ whose kernel is $\sum_i \phi_i(\mathbf{r}) n_i \phi_i^*(\mathbf{r}')$. The OEL equations are equivalent to $[\mathcal{G}, \hat{\rho}] = 0$. The corresponding time-dependent equations are $i\hbar\frac{\partial}{\partial t}\hat{\rho} = [\mathcal{G}(t), \hat{\rho}(t)]$. Dirac proved, for Hermitian $\mathcal{G}$, that the time-dependent equation $i\hbar\frac{\partial}{\partial t}\phi_i(\mathbf{r}t) = \mathcal{G}(\mathbf{r}t)\phi_i(\mathbf{r}t)$ implies that $\hat{\rho}(t)$ is idempotent. Hence $\hat{\rho}(t)$ corresponds to a normalized time-dependent reference state.

For $t \geq 0$, a weak perturbing potential $\Delta v(\mathbf{r}t)$ induces a screening potential such that $\Delta\mathcal{G} = \Delta v + \Delta\mathcal{G}_s$. The first-order perturbation equations are

$$i\hbar\frac{\partial}{\partial t}\Delta\hat{\rho} = [\mathcal{G}(0), \Delta\hat{\rho}] + [\Delta\mathcal{G}, \hat{\rho}(0)].$$

$\Delta\phi_i(t) = \sum_a (1 - n_a)\phi_a c_i^a(t)$ in this orbital representation. Hence the kernel of $\Delta\hat{\rho}$ is

$$\Delta\rho(\mathbf{r}, \mathbf{r}'; t) = \sum_i \sum_a n_i(1 - n_a)\big[\phi_a(\mathbf{r})c_i^a(t)\phi_i^*(\mathbf{r}') + \phi_i(\mathbf{r})c_i^a(t)^*\phi_a^*(\mathbf{r}')\big].$$

Using $(p|\mathcal{G}(0)|j) = \epsilon_p\delta_{pj}$ and $(j|\hat{\rho}(0)|q) = n_q\delta_{jq}$, in the basis of eigenfunctions of $\mathcal{G}(0)$

$$(p|[\mathcal{G}(0), \Delta\hat{\rho}]|q) = (\epsilon_p - \epsilon_q)(p|\Delta\hat{\rho}|q);$$
$$(p|[\Delta\mathcal{G}, \hat{\rho}(0)]|q) = (n_q - n_p)(p|\Delta\mathcal{G}|q).$$

The equation of motion implied for $c_i^a(t)$ is

$$i\hbar\dot{c}_i^a(t) = (\epsilon_a - \epsilon_i)c_i^a(t) + (n_i - n_a)(a|\Delta\mathcal{G}|i),$$

for $i \leq N < a$.

6.4.3 Definition of the response kernel

Because $\mathcal{G}$ itself is defined by an orbital functional derivative, the increment $\Delta\mathcal{G}_s$ is proportional to a functional second derivative. It is convenient to define a *response kernel* $\hat{f}$ such that $\Delta\phi_j(t)$ induces $\Delta\hat{v} = \sum_j n_j \sum_b (1 - n_b)[(j|\hat{f}|b)c_j^b(t) + c_j^{b*}(t)(b|\hat{f}|j)]$. If $\hat{v} = \frac{\delta F}{n_i\delta\phi_i^*}$, this equation defines a functional second derivative in the form $\hat{f}\phi_j(\mathbf{r}) = \frac{\delta\hat{v}}{n_j\delta\phi_j^*}$. In agreement with Dirac [79] and with the second-quantized Hamiltonian, the response kernel for Hartree and exchange energy functionals is $\hat{f}_h + \hat{f}_x = \bar{u} = \frac{1}{r_{12}}(1 - \hat{P}_{12})$.

The response kernel $\hat{f}_c$ is a linear operator such that $\hat{f}_c\phi_j = \frac{\delta\hat{v}_c}{n_j\delta\phi_j^*}$. Variations of unoccupied orbitals $\delta\phi_a$ $(N < a)$ in the functional E_c are induced by variations of occupied orbitals $\delta\phi_i$ $(i \leq N)$ through unitarity. The combined total response kernel is the linear operator $\hat{f}_h + \hat{f}_x + \hat{f}_c = \hat{f}_{hxc} = \bar{u} + \hat{f}_c$.

The equations of motion for the coefficients $c_i^a(t)$ are

$$i\hbar\dot{c}_i^a(t) = (\epsilon_a - \epsilon_i)c_i^a(t) + (n_i - n_a)\Big[(a|\Delta v(t)|i) + \sum_j\sum_b \{(aj|\hat{f}_{hxc}|ib)c_j^b(t) + (ab|\hat{f}_{hxc}|ij)c_j^b(t)^*\}\Big],$$

for indices $i, j \leq N < a, b$. Setting $\Delta v(\mathbf{r}t) = 2\Delta w(\mathbf{r})\Re(e^{-i\omega t})$ for complex frequency ω, the independent coefficients defined by

$$c_i^a(t) = X_i^a e^{-i\omega t} + Y_i^{a*}e^{i\omega^* t}$$

satisfy linear inhomogeneous equations

$$(\epsilon_a - \epsilon_i - \hbar\omega)X_i^a + (n_i - n_a)\sum_j\sum_b [(aj|\hat{f}_{hxc}|ib)X_j^b + (ab|\hat{f}_{hxc}|ij)Y_j^b] = -(n_i - n_a)(a|\Delta w|i),$$

$$(\epsilon_a - \epsilon_i + \hbar\omega)Y_i^a + (n_i - n_a)\sum_j\sum_b [(ij|\hat{f}_{hxc}|ab)X_j^b + (ib|\hat{f}_{hxc}|aj)Y_j^b] = -(n_i - n_a)(i|\Delta w|a).$$

In the exchange-only limit, these are the TDHF or RPA equations (Thouless [407], p. 89). If $\Delta w = 0$, excitation energies are values of $\hbar\omega$ for which the determinant of the residual homogeneous equations vanishes.

6.5 Reconciliation of N-electron theory and orbital models

The action integral defined in standard N-electron theory is

$$A_N[\Psi] = \int_{t_0}^{t_1} dt \left\langle i\hbar\frac{\partial}{\partial t} - \hat{H}\right\rangle = \int_{t_0}^{t_1} dt \left\{\left\langle i\hbar\frac{\partial}{\partial t}\right\rangle - E[\Psi]\right\},$$

where the notation $\langle\cdots\rangle$ denotes $(\Psi|\cdots|\Psi)/(\Psi|\Psi)$ for an arbitrarily normalized N-electron wave function $\Psi(t)$. The theory postulates that this action integral is stationary subject to fixed normalization. For comparison, the action integral of orbital functional theory is

$$A[\Phi] = \int_{t_0}^{t_1} dt \left\{\left(\Phi\left|i\hbar\frac{\partial}{\partial t}\right|\Phi\right) - E[\Phi]\right\}.$$

While $E[\Phi] = E[\Psi]$, the mean values of the time derivative are not equal. Even if $\hat{H}$ is independent of time, these values are constants $\sum_i n_i\epsilon_i$ and E, respectively. These constants can be reconciled by inserting a gauge potential $\gamma = E - \sum_i n_i\epsilon_i$ into A, determined by equating the time-dependent phase factors. This also preserves the normalization $(\Phi|\Psi) = (\Phi|\Phi) = 1$ for the time-dependent wave functions, so that the definition of correlation energy $E_c = (\Phi|\hat{H}|\Psi - \Phi)$ remains unchanged. When the Hamiltonian is time-dependent, $\gamma(t)$ must be determined so that $(\Phi_\gamma|i\hbar\frac{\partial}{\partial t}|\Phi_\gamma) = \langle i\hbar\frac{\partial}{\partial t}\rangle$, where Φ_γ incorporates the gauge-dependent phase factor. For any rule $\Psi \to \Phi$ applied for all $t_0 \leq t \leq t_1$, the time derivative terms in the action integrals are $\Theta_s = \int_{t_0}^{t_1} dt(\Phi|i\hbar\frac{\partial}{\partial t}|\Phi)$ and $\Theta = \int_{t_0}^{t_1} dt\langle i\hbar\frac{\partial}{\partial t}\rangle$. The present argument indicates that a gauge transformation can be introduced such that $\Theta_s - \Theta$ drops out of the variational formalism. This is invoked below to remove such a term from the equations of time-dependent density functional theory.

6.6 Time-dependent density functional theory (TDFT)

An extension of density functional theory to time-dependent external potentials has recently been derived [351, 155]. Hohenberg and Kohn [173] proved, for nondegenerate ground states, that the electronic density function $\rho(\mathbf{r})$ and the external potential function $v(\mathbf{r})$ mutually determine each other. Easily extended to spin-indexed density and potential, this implies that all properties of an N-electron ground state are functionals of the density function. A slightly more restrictive proof, requiring the existence of a Taylor expansion of the potential function in time, establishes an analogous 1–1 correspondence between the (spin-indexed) time-dependent density function $\rho(xt)$ and an external potential $v(xt)$ [351, 155]. This shows that all properties of the time-dependent state that evolves from a given initial state $\Psi(t_0)$ in the time interval $t_0 \leq t \leq t_1$ are functionals of $\rho(xt)$. This theorem is used to justify a time-dependent density functional theory (TDFT), based on a time-dependent generalization of the Kohn–Sham equations. These are modified time-dependent Schrödinger equations

$$i\hbar\frac{\partial}{\partial t}\phi_i(xt) = \{\hat{t} + w(xt)\}\phi_i(xt),$$

for each of N occupied orbital functions of a time-dependent model state Φ, itself a single Slater determinant. If a Fréchet functional derivative [26] of a density functional analogous to $E - T$ exists in this time-dependent theory, it defines an effective local potential $w(xt)$ in the orbital equations. If the assumed local effective potential is replaced by a more general linear operator or nonlocal potential, this formalism becomes identical to TOFT, as derived above, which reduces, for an exchange-only model, to the TDHF theory of Dirac [79].

Generalized Hohenberg–Kohn theorems are proved by developing the action integrals appropriate to the time-dependent theory [351, 155]. Assumptions can be minimized by carrying out a time-dependent version of the constrained-search construction of Levy [222, 201]. The variational condition that a certain action integral should be stationary subject to the constraint of specified density $\rho(xt)$ is implemented here by introducing the time-dependent external potential $v(xt)$ as a Lagrange-multiplier field for the density constraint. If these equations have a stationary solution, this establishes the property of "v-representability" of the specified density. An action integral is defined for arbitrarily normalized time-dependent N-electron trial functions $\Psi(t)$ by

$$B[\Psi(t)] = \int_{t_0}^{t_1} dt \left\langle i\hbar\frac{\partial}{\partial t} - \hat{T} - \hat{U}\right\rangle.$$

The N-electron Hamiltonian is $\hat{H} = \hat{T} + \hat{U} + \hat{V}$, where the three terms represent kinetic energy, interelectronic Coulomb interaction, and an external field, respectively. The variational condition that determines $\Psi[\rho](t)$ for $t_0 \leq t \leq t_1$ is

$$\delta\left(B[\Psi(t)] - \int_{t_0}^{t_1} dt \int d^3\mathbf{x}\, v(xt)[\rho_\Psi(xt) - \rho(xt)]\right) = 0,$$

for variations of Ψ within the N-electron Hilbert space. Given a particular field $v(xt)$, the stationary condition determines $\Psi_v(t)$ through the time-dependent Schrödinger equations, but in general the density $\rho_v(xt)$ computed from this wave function differs from the specified $\rho(xt)$. A Lagrange-multiplier field $v_\rho(xt)$ is to be chosen such that $\rho_v(xt) = \rho(xt)$. Since $\int v\rho = \int_{t_0}^{t_1} dt \int d^3\mathbf{x}\, v(xt)\rho(xt)$ is constant for given v, $A_v[\rho] = B[\rho] + \int v\rho$ is defined as a functional of ρ, where $B[\rho] = B[\Psi_v]$, evaluated for $v = v_\rho$. This construction must succeed whenever the time-dependent Schrödinger equation has a solution in the interval $t_0 \leq t \leq t_1$ with $\Psi(t_0) = \Psi_0$, and if the Lagrange multiplier field $v_\rho(xt)$ exists. This latter condition is "v-representability". These conditions seem very plausible for variations about well-defined physical solutions, but there is no rigorous theory giving general uniqueness and existence conditions. When this construction succeeds, it determines a functional $A_v[\rho]$ that has a stationary value when $\rho(xt)$ is determined by a solution of the time-dependent Schrödinger equation defined by the external potential function $v_\rho(xt)$. Uniqueness follows from uniqueness of the solution of this time-dependent equation [351], within a class of equivalent solutions corresponding to gauge transformations. The trial functional B is "universal" in the sense that it depends only on $\hat{H} - \hat{V}$, the same for all external potentials. However, the variational condition implies different results for different initial conditions, $\Psi(t_0) = \Psi_0$.

Defining the functional $\Theta = \int_{t_0}^{t_1} dt \langle i\hbar\frac{\partial}{\partial t}\rangle$, then $A_v = \Theta - \Xi$, where $\Xi = \int_{t_0}^{t_1} dt\, E_v(t)$. Using $E_v(t) = \langle\hat{H}\rangle = \langle\hat{f} + \hat{V}\rangle = F + V$, stationary A_v implies

Euler–Lagrange equations of the form

$$\frac{\delta\Theta}{\delta\rho} = \frac{\delta F}{\delta\rho} + v(xt),$$

if the indicated density functional derivatives exist as Fréchet derivatives, equal to multiplicative local potential functions.

Applying the same argument to a noninteracting N-electron system with the same density defines an action integral

$$B_s[\Phi(t)] = \int_{t_0}^{t_1} dt \left(\Phi\middle| i\hbar\frac{\partial}{\partial t} - \hat{T}\middle|\Phi\right),$$

where the N-electron trial functions are model functions Φ, in the form of single Slater determinants. A Lagrange-multiplier field $w_\rho(xt)$ is determined such that the solution of the variational condition

$$\delta\left(B_s[\Phi(t)] - \int_{t_0}^{t_1} dt \int d^3\mathbf{x}\, w(xt)[\rho_\Phi(xt) - \rho(xt)]\right) = 0,$$

determines a density equal to the specified $\rho(xt)$ when $w(xt) = w_\rho(xt)$. "Noninteracting v-representability" is established by any solution of these equations. This defines a "universal" density functional $B_s[\rho]$ when evaluated for ρ and w_ρ. For fixed $w(xt)$ but variable ρ, the functional $A_{sw}[\rho] = B_s[\rho] + \int w\rho$ is stationary when $\rho = \rho_w$, the density obtained by solving the noninteracting time-dependent Schrödinger equation in which the time-dependent potential is $w_\rho = w$.

Defining the functional $\Theta_s = \int_{t_0}^{t_1} dt(\Phi|i\hbar\frac{\partial}{\partial t}|\Phi)$, then $A_{sw} = \Theta_s - \Xi_s$, where $\Xi_s = \int_{t_0}^{t_1} dt(\Phi|\hat{T} + \hat{W}|\Phi)$. Stationary A_{sw} implies Euler–Lagrange equations of the form

$$\frac{\delta\Theta_s}{\delta\rho} = \frac{\delta T_s}{\delta\rho} + w(xt),$$

if the indicated density functional derivatives exist as Fréchet derivatives.

In the noninteracting problem both required functional derivatives can be evaluated from explicit orbital functional derivatives. In detail, $\frac{\delta\Theta_s}{n_i\delta\phi_i^*} = i\hbar\frac{\partial}{\partial t}\phi_i$ and $\frac{\delta T_s}{n_i\delta\phi_i^*} = \hat{t}\phi_i$. For variations of the partial densities $\rho_i = \phi_i^*\phi_i$, this implies Gâteaux functional derivatives [26, 102] $\frac{\delta\Theta_s}{n_i\delta\rho_i}\phi_i = i\hbar\frac{\partial}{\partial t}\phi_i$ and $\frac{\delta T_s}{n_i\delta\rho_i}\phi_i = \hat{t}\phi_i$. The implied orbital Euler–Lagrange equations are the time-dependent Schrödinger equations for the occupied orbitals $i \leq N$,

$$i\hbar\frac{\partial}{\partial t}\phi_i(xt) = (\hat{t} + w(xt))\phi_i(xt).$$

Thus for noninteracting electrons, density functional analysis reproduces the usual Schrödinger equations, but a theory based solely on densities (a time-dependent

Thomas–Fermi theory) requires a Fréchet derivative for the kinetic energy, and cannot exist for more than two electrons [288].

For interacting electrons, the required functional derivatives cannot be evaluated explicitly. In the TDFT derivation [155], the density functional derivative of the kinetic energy term in Ξ is treated correctly as equivalent to the Schrödinger operator $\hat{t}$. Nonetheless, the functional derivative of the exchange-correlation functional is assumed without proof to be a Fréchet derivative, equal to a time-dependent local potential function $v_{xc}(xt)$. It is argued above that the difference functional $\Theta - \Theta_s$ can consistently be set to zero by a gauge transformation. Taking this into account, and assuming that v_{xc} exists, both interacting and noninteracting equations have the same form, and by construction produce the same density function $\rho(xt)$. It can then by inferred that the local potentials in these equations are equal, up to a gauge potential $\gamma(t)$. Then the Kohn–Sham potential could be decomposed into $w(xt) = v(xt) + v_h(xt) + v_{xc}(xt)$. Hence if a Fréchet derivative v_{xc} did exist, the correct interacting $\rho(xt)$ would be obtained by solving the implied time-dependent Kohn–Sham equation.

This proposition has been tested in the exact-exchange limit of the implied linear-response theory [329]. The TDFT exchange response kernel disagrees qualitatively with the corresponding expression in Dirac's TDHF theory [79, 289]. This can be taken as evidence that an exact local exchange potential does not exist in the form of a Fréchet derivative of the exchange energy functional in TDFT theory.

6.7 Excitation energies and energy gaps

Time-dependent OFT implies matrix equations for excitation energies $\hbar\omega$,

$$(\epsilon_a - \epsilon_i - \hbar\omega)X_i^a + (n_i - n_a)\sum_j\sum_b\left[(aj|\hat{f}_{hxc}|ib)X_j^b + (ab|\hat{f}_{hxc}|ij)Y_j^b\right] = 0,$$

$$(\epsilon_a - \epsilon_i + \hbar\omega)Y_i^a + (n_i - n_a)\sum_j\sum_b\left[(ij|\hat{f}_{hxc}|ab)X_j^b + (ib|\hat{f}_{hxc}|aj)Y_j^b\right] = 0.$$

The simplest internally consistent approximation to an excitation energy is obtained by limiting the summation to the diagonal term $j, b = i, a$. The second line vanishes because of antisymmetry, $(ii|\hat{f}_{hxc}|aa) \equiv 0$, and the first line reduces to a single equation,

$$\{\epsilon_a - \epsilon_i - \hbar\omega - (ai|\hat{f}_{hxc}|ai)\}X_i^a = 0.$$

Neglecting correlation response, this implies a well-known formula for zeroeth-order hole–particle excitation energies [261, 149], $\hbar\omega = \epsilon_a - \epsilon_i - (ai|\bar{u}|ai)$. The two-electron integral here depends strongly on orbital localization. Since the lower state is stationary, this energy difference should be minimized to represent a stationary excited state. This can be done by separate localization transformations of occupied and unoccupied orbitals. This provides a mechanism to reduce Hartree–Fock band gaps, which are systematically too large. The correlation response kernel has the physical effect of screening the interelectronic Hartree and exchange terms. This implies direct effects on energy gaps, evident in the TOFT formalism.

III

Continuum states and scattering theory

This part extends quantum variational theory to continuum states. In particular, variational principles are developed for wave function continuity at specified energy, which is the usual context of scattering theory. Chapter 7, concerned with multiple scattering theory, lies somewhere between the theory of bound states and true scattering theory. Formalism appropriate to the latter is adapted to computing the electronic structure of large molecules and periodic solids, whose energy levels are determined by consistency conditions for wave function continuity. A variational formalism is derived for energy linearization. Chapter 8 develops variational principles and methods suitable for the true continuum problem of electron scattering at specified energy. Chapter 9 presents methodology, some very recent, that allows rotational and vibrational effects in electron–molecule scattering to be treated as a practicable extension of fixed-nuclei variational theory.

7

Multiple scattering theory for molecules and solids

The principal references for this chapter are:

[52] Butler, W.H. and Zhang, X.-G. (1991). Accuracy and convergence properties of multiple-scattering theory in three dimensions, *Phys. Rev. B* **44**, 969–983.
[113] Ferreira, L.G. and Leite, J.R. (1978). General formulation of the variational cellular method for molecules and crystals, *Phys. Rev. A* **18**, 335–343.
[148] Gonis, A. and Butler, W.H. (2000). *Multiple Scattering in Solids* (Springer, New York).
[203] Kohn, W. (1952). Variational methods for periodic lattices, *Phys. Rev.* **87**, 472–481.
[204] Kohn, W. and Rostoker, N. (1954). Solution of the Schrödinger equation in periodic lattices with an application to metallic lithium, *Phys. Rev.* **94**, 1111–1120.
[277] Nesbet, R.K. (1990). Full-potential multiple scattering theory, *Phys. Rev. B* **41**, 4948–4952.
[279] Nesbet, R.K. (1992). Variational principles for full-potential multiple scattering theory, *Mat. Res. Symp. Proc.* **253**, 153–158.
[359] Schlosser, H. and Marcus, P. (1963). Composite wave variational method for solution of the energy-band problem in solids, *Phys. Rev.* **131**, 2529–2546.
[384] Skriver, H.L. (1984). *The LMTO Method* (Springer-Verlag, Berlin).
[432] Williams, A.R. and Morgan, J. van W. (1974). Multiple scattering by non-muffin-tin potentials: general formulation, *J. Phys. C* **7**, 37–60.

For direct N-electron variational methods, the computational effort increases so rapidly with increasing N that alternative simplified methods must be used for calculations of the electronic structure of large molecules and solids. Especially for calculations of the electronic energy levels of solids (energy-band structure), the methodology of choice is that of independent-electron models, usually in the framework of density functional theory [189, 321, 90]. When restricted to local potentials, as in the local-density approximation (LDA), this is a valid variational theory for any N-electron system. It can readily be applied to heavy atoms by relativistic or semirelativistic modification of the kinetic energy operator in the orbital Kohn–Sham equations [229, 384].

Solution of the one-electron Schrödinger or modified Dirac equation for a system of many atoms is still a difficult computational task. This has been simplified and made tractable by the methodology of multiple scattering theory (MST), recently reviewed by Gonis and Butler [148]. These authors concentrate on methods in which orbital energy levels are identified by searching for roots of a secular determinant appropriate to the Kohn–Sham equations. Although MST uses the formalism of scattering theory, it is used in practice as a bound-state method, computing energy levels by varying a specified energy until continuity conditions are satisfied. In particular, the electronic energy bands of regular periodic solids for specified translational quantum numbers (momentum vectors in a reduced Brillouin zone [384, 148]) are defined as a discrete set of energy values that cause a secular determinant to vanish. True continuum scattering theory is discussed in the following chapter.

At the cost of introducing an additional level of approximation, root-search methodology can be replaced within some fixed interval of electronic energy levels (an energy *panel* or *window*) by an energy-linearized method which requires solution of a matrix eigenvalue problem for energy eigenvalues within the panel. Because of the very substantial relative gain in efficiency, such linearized methods [384] are by far the most widely used in energy-band calculations. Both root-search and energy-linearized MST methods require iteration to self-consistency, converging when the computed density function is sufficiently close to that used to construct the potential function in the current iteration. This requires accurate solution of the Poisson equation determined by the input density function, which is also simplified by an adaptation of MST [148].

In addition to the model approximation implied by a restriction to local exchange and correlation potentials, the methodology has often been further simplified by the *muffin-tin* model. In this model, each atom of a molecule or solid is surrounded by a muffin-tin sphere, of sufficiently small radius that adjacent spheres do not overlap. The local potential function is spherically averaged within each such sphere, and set to a constant value (the muffin-tin zero) in the interstitial volume between these spheres. This mathematical model greatly simplifies the required computational effort. It is physically justifiable for metals, since the interstitial region is described in terms of electronic free wave functions, but it is counterintuitive for atoms in molecules, since electronic exchange and correlation in chemical bonds must be treated accurately. Earlier work restricted to the muffin-tin model has been supplanted by full-potential methodology in recent years, based on the full local potential of the Kohn–Sham model, without simplifying approximations.

The most natural geometrical framework for full-potential methodology is that of space-filling polyhedral atomic cells, mathematically a Voronoi lattice construction in three dimensions. Space is subdivided by perpendicular bisector planes of the internuclear coordinate vectors. This lattice construction corresponds to the Wigner–Seitz [429] cellular model. For open lattice structures (e.g. the diamond lattice), it is common practice to introduce empty cells, with no enclosed atomic nucleus, to provide a more uniform space-filling structure. Variational principles and methods appropriate to such space-filling cellular lattices will be discussed here. The muffin-tin model is a simplified special case in this general theory. A closely related approximation, the atomic sphere model [384], extends the spherically averaged local potential of the muffin-tin sphere throughout a concentric sphere whose volume equals that of the Wigner–Seitz polyhedral atomic cell. This can be extended to the full enclosing sphere of each polyhedral atomic cell, which greatly simplifies the computation of atomic basis functions required in full-potential MST. These spherically averaged potentials can be considered as initial approximations to be used during self-consistency iterations, to be modified following convergence by nonspherical potential terms to be treated as perturbations in the context of linear response theory. Nonlocal corrections to an initial local-potential model may also be included in principle within such a linear-response theory.

7.1 Full-potential multiple scattering theory

Since the principal applications of MST are in solid-state theory, which traditionally uses Rydberg rather than Hartree units of energy, this convention will be followed in the present chapter. Restricting the formalism to local potential functions as in Kohn–Sham theory, the Schrödinger equation for an electron of positive energy is

$$(\nabla^2 + \kappa^2)\psi(\mathbf{x}) = v(\mathbf{x})\psi(\mathbf{x}),$$

where the one-electron energy is defined by $\epsilon = \kappa^2$ (Ryd). The momentum κ is a parameter that characterizes electronic energies above a zero value assigned to the lowest energy of an energy band. Since this Schrödinger equation is formally a modified Helmholtz equation, it is convenient to introduce a Helmholtz Green function defined such that

$$(\nabla^2 + \kappa^2)G_0(\mathbf{x}, \mathbf{x}') = \delta(\mathbf{x}, \mathbf{x}').$$

The Green function must satisfy boundary conditions at large distances consistent with the wave function ψ. The Schrödinger equation can be replaced by an equivalent Lippmann–Schwinger integral equation

$$\psi(\mathbf{x}) = \chi(\mathbf{x}) + \int_{\Re^3} G_0(\mathbf{x}, \mathbf{x}')v(\mathbf{x}')\psi(\mathbf{x}')\, d^3\mathbf{x}',$$

where χ is a solution of the homogeneous Helmholtz equation. In an operator notation this is $\psi = \chi + G_0 v \psi$. For a solid or molecule, coordinate space $\Re^3$ can be subdivided into space-filling cells, each containing no more than one atomic nucleus. A spherical polar coordinate system in each cell is centered about this Coulombic singularity if it exists. The closed surface of cell τ_μ is denoted by σ_μ.

7.1.1 Definitions

Solid-harmonic solutions J_L^μ and N_L^μ of the homogeneous Helmholtz equation in cell μ are products of spherical Bessel functions and spherical harmonics. Specific functional forms for the regular and irregular solid harmonics, respectively, are [188]

$$J_L = \kappa^{\frac{1}{2}} j_\ell(\kappa r) Y_{\ell,m}(\theta, \phi),$$
$$N_L = -\kappa^{\frac{1}{2}} n_\ell(\kappa r) Y_{\ell,m}(\theta, \phi).$$

The standard angular momentum indices ℓ, m, and an implicit spin index, are here summarized by a single collective index L. Relative normalization is such that $(N_L|W_\sigma|J_{L'}) = \delta_{L,L'}$, expressed in terms of Wronskian integrals over surface σ defined by

$$(\phi_1|W_\sigma|\phi_2) = \int_\sigma [\phi_1^* \nabla_n \phi_2 - (\nabla_n \phi_1)^* \phi_2]\, d\sigma.$$

Here ∇_n denotes the outward normal gradient on σ. It can easily be verified from the defining differential equation that the Helmholtz Green function is given by the expansion

$$G_0(\mathbf{r}, \mathbf{r}') = -\sum_L J_L(\mathbf{r}) N_L^*(\mathbf{r}'), \qquad r < r',$$

in any specified atomic cell [188]. The definition of irregular solid harmonics N_L here establishes boundary conditions appropriate to the *principal value* Green function in scattering theory.

7.1.2 Two-center expansion

MST is based on expanding the total wave function in the local coordinate system of each atomic cell. The Helmholtz Green function has an explicit expansion of this sort. Given two distinct cells τ_μ and τ_ν, with coordinate origins $\mathbf{X}_\mu$ and $\mathbf{X}_\nu$, respectively, global coordinates take the local form $\mathbf{x} = \mathbf{X}_\mu + \mathbf{r}$ in τ_μ and $\mathbf{x}' = \mathbf{X}_\nu + \mathbf{r}'$ in τ_ν. The Green function is required in the local coordinates $\mathbf{r}, \mathbf{r}'$. Now $N_L^\mu(\mathbf{x}' - \mathbf{X}_\mu)$ is a regular solution of the Helmholtz equation in a sphere of radius $|\mathbf{X}_\mu - \mathbf{X}_\nu|$ about the origin of cell $\nu \neq \mu$. This determines a local expansion of the form

$$N_L^{\mu}(\mathbf{r}' + \mathbf{X}_\nu - \mathbf{X}_\mu) = -\sum_{L'} J_{L'}^{\nu}(\mathbf{r}') g_{L',L}^{\nu\mu},$$

for $r' < |\mathbf{X}_\mu - \mathbf{X}_\nu|$. On substituting this into the one-center expansion,

$$G_0(\mathbf{X}_\mu + \mathbf{r}, \mathbf{X}_\nu + \mathbf{r}') = \sum_L \sum_{L'} J_L^{\mu}(\mathbf{r}) g_{L,L'}^{\mu\nu} J_{L'}^{\nu *}(\mathbf{r}'),$$

for $r, r' < |\mathbf{X}_\mu - \mathbf{X}_\nu|$. The Hermitian matrix $g_{L,L'}^{\mu\nu}$ defines *structure constants* for the two-center expansion of the Green function in locally regular solid harmonics. Through the Lippmann–Schwinger equation, matrix g propagates a given wave function in one cell into incremental wave functions in the other cell. MST collects the total incremental wave function in each cell as a multicenter expansion, to establish a linear consistency condition for a molecule or solid. For a regular periodic lattice, contributions from outside an elementary polyatomic translational cell are multiplied by translational phase factors and summed over an infinite space-lattice.

7.1.3 Angular momentum representation

An eigenfunction ψ of the Schrödinger equation satisfies the homogeneous integral equation $\psi = \int_{\Re^3} G_0 v \psi$. In cell τ this defines a locally regular solution of the Helmholtz equation

$$\chi = \psi - \int_\tau G_0 v \psi = \int_{\Re^3 - \tau} G_0 v \psi,$$

consistent with the Lippmann–Schwinger equation.

The function χ can be represented in τ by a sum of regular solid harmonic functions,

$$\chi(\mathbf{r}) = \sum_L J_L(\mathbf{r}) C_L.$$

Similarly, because ψ is regular at the local coordinate origin, it can be represented at this origin by

$$\psi(\mathbf{r}) = \sum_L J_L(\mathbf{r}) c_L.$$

From the Lippmann–Schwinger equation and the expansion of G_0 in some specified cell τ_μ, the coefficients in these expansions must be related by $C_L^{\mu} = c_L^{\mu} + \int_{\tau_\mu} N_L^* v \psi$. Consistency conditions are derived by considering the alternative expansion of χ in cell τ_μ,

$$\chi(\mathbf{r}) = \sum_{\nu\neq\mu} \int_{\tau_\nu} G_0 v\psi.$$

From the two-center expansion of the Green function, this equation implies

$$\chi(\mathbf{r}) = -\sum_L J_L(\mathbf{r}) \sum_{L'} \sum_{\nu\neq\mu} g^{\mu\nu}_{L,L'} S^\nu_{L'},$$

where $S^\nu_L = -\int_{\tau_\nu} J^*_L v\psi$. On equating both expressions for χ, and matching coefficients in the regular expansion about the coordinate origin in cell τ_μ,

$$C^\mu_L + \sum_{L'} \sum_{\nu\neq\mu} g^{\mu\nu}_{L,L'} S^\nu_{L'} = 0.$$

This derivation implies that these equations are valid under the condition that no adjacent nucleus should lie within the enclosing sphere of a given local cell.

The wave function ψ can be expanded as a linear combination of basis functions defined separately in each atomic cell. It is convenient to define specific basis functions $\phi_L(\mathbf{r})$ constructed by solving the local Schrödinger equation at a specified orbital energy and matching to $J_L(\mathbf{r})$ at the cell origin. Given these primitive basis functions, in one-to-one correspondence with locally defined regular solid harmonics, the local expansion is $\psi = \sum_L \phi_L \gamma_L$. If the local potential is nonspherical, each basis function ϕ_L becomes a sum of spherical-harmonic components as it is integrated outward from the local cell origin. Nevertheless, it is completely characterized by the single index L, and by a spin index that is assumed in the notation and discussion here. In this basis,

$$\sum_{\nu,L',L''} (\delta^{\mu\nu}_{L,L''} C^\nu_{L'',L'} + g^{\mu\nu}_{L,L''} S^\nu_{L'',L'}) \gamma^\nu_{L'} = 0,$$

where

$$C_{L,L'} = \delta_{L,L'} + \int_\tau N^*_L v\phi_{L'}, \qquad S_{L,L'} = -\int_\tau J^*_L v\phi_{L'}. \tag{7.1}$$

If the matrix C is not singular, which requires the number of basis functions to match the number of solid harmonics used to expand the Green function, a local t-matrix is defined by $t = -SC^{-1}$. The consistency condition expressed above in terms of C and S matrices then reduces to the simple matrix expression

$$(t^{-1} - g)S\gamma = 0.$$

This is the fundamental equation of multiple scattering theory. It has the remarkable property of concentrating effects of the local potential function into the

t-matrix, separately determined in each local atomic cell, while the geometry of the polyatomic superstructure is entirely characterized by the matrix g of structure constants. For a regular periodic solid, the matrices here are summed over equivalent translational cells, with coefficient phase factors that depend on an effective momentum vector $\mathbf{k}$, to give

$$\{t^{-1}(\epsilon) - g(\epsilon;\mathbf{k})\}S(\epsilon)\gamma(\epsilon;\mathbf{k}) = 0.$$

The Korringa–Kohn–Rostoker (KKR) method [207, 204] is implemented by searching for the zeroes of the secular determinant $\det(t^{-1} - g)$, or $\det(I - tg)$ if t is singular.

The single-site matrix t corresponds to a scattering operator $\hat{t}$ defined by $\hat{t}\chi_L = v\phi_L$ for any primitive basis orbital ϕ_L in a given atomic cell. Here $\chi_L = \phi_L - \int_\tau G_0 v\phi_L = \sum_{L'} J_{L'}C_{L',L}$. From this definition it follows that $\hat{t}J_L = v\sum_{L'}\phi_{L'}(C^{-1})_{L',L}$. The t-matrix is

$$(J_L|\hat{t}|J_{L'}) = \sum_{L''}(J_L|v|\phi_{L''})(C^{-1})_{L'',L'}.$$

This implies that $t = -SC^{-1}$, since $(J_L|v|\phi_{L'}) = -S_{L,L'}$.

7.1.4 The surface matching theorem

Consider an atomic cell τ with closed boundary surface σ. Given a function ψ and its normal gradient $\nabla_n\psi$ on the surface σ, a unique solution of the Helmholtz equation is determined by either inward or outward integration. Because this global function is completely determined on σ, it in general will be singular at the coordinate origin of cell τ and unbounded at infinite distance from this center. Inward integration in general gives a singular function, which includes irregular functions N. For a function that is regular at the origin, either ψ (classical Dirichlet problem), or $\nabla_n\psi$ (Neumann problem) can be specified on σ, but not both [74, 255]. An expansion valid on surface σ will be established here, determined by both the value and normal gradient of ψ. Consistency of the surface Wronskian integrals implies that coefficients of an expansion in both J_L and N_L must be given by

$$\psi(\mathbf{r}_\sigma) \equiv_\sigma \sum_L [J_L(N_L|W_\sigma|\psi) - N_L(J_L|W_\sigma|\psi)]$$

if the sums converge. The notation here denotes matching of both function value and normal gradient. Even if the separate series in functions J_L and N_L do not converge, it can be shown that well-defined functions exist, providing summation formulas for the possibly divergent series. This justifies formal operations in MST that depend on Wronskian integrals of this expansion.

An interior function $\chi(\mathbf{r}) = -(G_0(\mathbf{r}, \mathbf{r}_\sigma)|W_\sigma|\psi)$ is defined that is regular at the cell origin. The one-center expansion of the Green function converges in the largest sphere enclosed by surface σ, sphere S_0 of radius r_0. This determines a series expansion $\chi(\mathbf{r}) = \sum_L J_L(N_L|W_\sigma|\psi)$, valid for $r < r_0$. Similarly, an exterior function $\eta(\mathbf{r}) = (G_0(\mathbf{r}, \mathbf{r}_\sigma)|W_\sigma|\psi)$ is defined that is bounded at infinite separation. Again using the convergent expansion of the Green function, but in the exterior region outside an enclosing sphere S_1 of radius r_1, $\eta(\mathbf{r}) = -\sum_L N_L(J_L|W_\sigma|\psi)$ for $r > r_1$.

The auxiliary functions χ and η have finite values on σ because the surface area element removes the singularity of the Green function in their defining surface integrals. Both functions are defined throughout the region $r_0 \leq r \leq r_1$. The boundary conditions determine unique solutions of the Helmholtz equation for integration of χ outward from S_0 and of η inward from S_1. Similarly, the function ψ given on σ can be extended throughout this region by integrating the Helmholtz equation both inwards to S_0 and outwards to S_1. Green's theorem implies for such solutions of the Helmholtz equation that Wronskian surface integrals are conserved on nested closed surfaces. Hence the difference function $\psi - \chi - \eta$ has vanishing Wronskian integrals with all regular and irregular solid harmonics on both spherical surfaces S_0 and S_1. It follows from Green's theorem that this difference function must vanish identically in the region between these two surfaces, and in particular on surface σ. This establishes the *surface matching theorem* [280],

$$\lim_{\mathbf{r}\to\mathbf{r}_{\sigma-}} \chi + \lim_{\mathbf{r}\to\mathbf{r}_{\sigma+}} \eta \equiv_\sigma \psi(\mathbf{r}_\sigma).$$

The series expansion of χ may diverge in the interior *moon region* [148], between the enclosed sphere S_0 and σ, while the corresponding series for η may diverge between σ and S_1. Nonetheless, these functions are well defined by Wronskian integrals of the Green function, independently of such expansions. The implied surface integrals are properties of the integrated functions, and justify formal use of the surface series expansion in boundary matching conditions based on these Wronskian integrals. If applied to basis functions that are solutions of the same Schrödinger equation in τ, for which $(\phi_i|W_\sigma|\phi_j) = 0$, surface expansion of either function implies

$$\sum_L (\phi_i|W_\sigma|J_L)(N_L|W_\sigma|\phi_j) - \sum_L (\phi_i|W_\sigma|N_L)(J_L|W_\sigma|\phi_j) = 0.$$

7.1.5 Surface integral formalism

Formalism from scattering theory, to be developed in a later chapter, shows that a function ψ and its normal gradient on a closed surface σ can be matched to a linear combination of regular internal solutions of a Schrödinger equation if and only if

$$\psi(\mathbf{r}_\sigma) = \int_\sigma \Re(\epsilon; \mathbf{r}_\sigma, \mathbf{r}'_\sigma)\nabla_n \psi(\mathbf{r}'_\sigma)\, d\mathbf{r}'_\sigma.$$

Here $\Re$ is an Hermitian linear integral operator over σ that can be constructed variationally from basis solutions $\{\phi_i\}$ of the Schrödinger equation that are regular in the enclosed volume. The functions ϕ_i do not have to be defined outside the enclosing surface and in fact must not be constrained by a fixed boundary condition on this surface [270]. This is equivalent to the Wronskian integral condition $(\phi_i|W_\sigma|\psi) = 0$, for all such ϕ_i. When applied to particular solutions of the form $\psi = J - Nt$ on σ,

$$(\phi_i|W_\sigma|J_L) - \sum_{L'} (\phi_i|W_\sigma|N_{L'})t_{L',L} = 0.$$

This equation determines the local t-matrix. Any local basis function ϕ_i has the formal expansion on σ

$$\phi_i \equiv_\sigma \sum_L [J_L(N_L|W_\sigma|\phi_i) - N_L(J_L|W_\sigma|\phi_i)].$$

Canonical basis orbitals are defined here such that $\phi_L \to J_L$ at the local cell origin. Introducing a matrix notation for surface Wronskian integrals,

$$C_{L,L'} = (N_L|W_\sigma|\phi_{L'}), \qquad S_{L,L'} = -(J_L|W_\sigma|\phi_{L'}), \tag{7.2}$$

the formal expansion of a primitive basis orbital ϕ_L on σ is

$$\phi_L \equiv_\sigma \sum_{L'} (J_{L'}C_{L',L} + N_{L'}S_{L',L}).$$

The surface Wronskian integrals C and S defined here are identical with the volume integrals defined in Eqs. (7.1). Using $(\nabla^2 + \kappa^2)\phi_L = v\phi_L$, integration by parts of the MST volume integrals, Eqs. (7.1), gives equivalent surface integrals, verifying Eqs. (7.2),

$$S_{L,L'} = -\int_\tau J_L^*(\nabla^2 + \kappa^2)\phi_{L'} = -(J_L|W_\sigma|\phi_{L'}),$$

$$C_{L,L'} = \delta_{L,L'} + \int_\tau N_L^*(\nabla^2 + \kappa^2)\phi_{L'} = (N_L|W_\sigma|\phi_{L'}).$$

7.1.6 Muffin-tin orbitals and atomic-cell orbitals

Deriving an energy-linearized version of MST, Andersen [12, 9] introduced muffin-tin orbital (MTO) basis functions. These functions have the form $\phi - \chi$ inside S_0, the enclosed sphere of an atomic cell, and η outside. Here ϕ is a local solution of the Schrödinger equation, while χ and η are sums of regular and irregular solid harmonics, respectively. The matching on S_0 is exact in the muffin-tin model, so that these functions provide a basis set suitable for variational calculations. In practice, because the long-range η functions are not orthogonal to inner shell functions in displaced atoms [193], they are expanded in the energy derivatives $\dot{\phi}$ of the basis functions ϕ. The resulting linearized muffin-tin orbital (LMTO) method [13, 9, 384] solves the Schrödinger secular equation in an MTO basis, constructed using KKR/MST structure constants.

The surface matching theorem makes it possible to generalize the idea of muffin-tin orbitals to a nonspherical Wigner–Seitz cell τ. Each local basis orbital is represented as $\phi \equiv_\sigma \chi + \eta$ on the cell surface σ, where χ and η are the auxiliary functions defined by the surface matching theorem. An atomic-cell orbital (ACO) is defined as the function $\phi - \chi$, regular inside τ. By construction, the smooth continuation of this ACO outside τ is the function η. The specific functional forms are

$$\chi_L = \sum_{L'} J_{L'} C_{L',L}, \qquad \eta_L = \sum_{L'} N_{L'} S_{L',L}.$$

Either MTO or ACO functions are valid as basis functions for expanding a global wave function ψ in all atomic cells. By construction, they are regular in τ, smooth at σ, and bounded outside. When the matrix C is nonsingular, modified canonical basis functions can be defined such that

$$\hat{\phi}_L = \sum_{L'} \phi_{L'} (C^{-1})_{L',L},$$

so that the auxiliary regular function is $\hat{\chi}_L = J_L$. The corresponding ACO function takes the especially simple form $\hat{\phi}_L - J_L$, matched on the cell surface σ to $\hat{\eta}_L = -\sum_{L'} N_{L'} t_{L',L}$.

7.1.7 Tail cancellation and the global matching function

As discussed by Andersen [9, 10] for muffin-tin orbitals, the locally regular components χ defined in each muffin-tin sphere are cancelled exactly if expansion coefficients satisfy the MST equations (the *tail-cancellation* condition) [9, 384]. The standard MST equations for space-filling cells can be derived by shrinking the interstitial volume to a *honeycomb lattice* surface that forms a common boundary for all cells. The wave function and its normal gradient evaluated on this honeycomb interface define a *global matching function* $\xi(\sigma)$.

In any cell $\tau = \tau_\mu$, with surface $\sigma = \sigma_\mu$, any solution ψ of the Lippmann–Schwinger equation defines auxiliary functions $\chi^{in} = \psi - \int_\tau G_0 v\psi$ and $\chi^{out} = \int_{\Re^3-\tau} G_0 v\psi$, which are equal by construction. After integration by parts,

$$\chi^{in} = -(G_0|W_\mu|\psi) = \sum_L J_L^\mu (N_L^\mu|W_\mu|\psi).$$

The notation W_μ defines a Wronskian integral over σ_μ. By the surface matching theorem, $\chi^{in} = \chi$, the interior component of ψ. Since ψ is a solution of the Lippmann–Schwinger equation, this implies $\chi^{out} = \chi^{in}$ when evaluated in the interior of τ_μ. This is a particular statement of the tail-cancellation condition. To show this in detail, after integration by parts

$$\chi^{out} = \sum_{\nu\neq\mu}(G_0|W_\nu|\psi) = \sum_{\nu\neq\mu}\sum_{L',L} J_{L'}^\mu g_{L',L}^{\mu\nu}(J_L^\nu|W_\nu|\psi),$$

which determines the standard MST equations.

On substituting the expansion $N_L^\nu = -\sum_{L'} J_{L'}^\mu g_{L',L}^{\mu\nu}$,

$$\chi^{out} = -\sum_{\nu\neq\mu}\sum_L N_L^\nu (J_L^\nu|W_\nu|\psi).$$

For comparison, $\eta = -\sum_L N_L^\mu (J_L^\mu|W_\mu|\psi)$ is the exterior component of ψ on $\sigma = \sigma_\mu$. Since $\chi^{out} \equiv_\sigma \chi$ on σ and $\xi \equiv_\sigma \chi + \eta$, this determines ξ everywhere on the global matching surface,

$$\xi \equiv_\sigma \sum_\nu \eta^\nu = \sum_\nu\sum_L N_L^\nu \beta_L^\nu, \tag{7.3}$$

where $\beta_L^\nu = -(J_L^\nu|W_\nu|\psi)$. If the sum here diverges, specific integrated functions η^ν should be used.

All Wronskian integrals of $\psi - \xi$ should vanish on the honeycomb lattice. Given the local expansion $\psi = \sum_L \phi_L^\mu \gamma_L^\mu$ in any particular cell τ_μ, $(J_L^\mu|W_\mu|\psi - \xi) = 0$ determines $\beta_L^\mu = \sum_{L'} S_{L,L'}^\mu \gamma_{L'}^\mu$. $(N_L^\mu|W_\mu|\psi - \xi) = 0$ implies that

$$\sum_{\nu,L',L''} (\delta_{L,L''}^{\mu\nu} C_{L'',L'}^\nu + g_{L,L''}^{\mu\nu} S_{L'',L'}^\nu)\gamma_{L'}^\nu = 0. \tag{7.4}$$

These are the standard MST equations, in the form $(t^{-1} - g)S\gamma = 0$, where

$$t = -SC^{-1} = (J|W_\sigma|\phi)(N|W_\sigma|\phi)^{-1}. \tag{7.5}$$

7.1.8 Implementation of the theory

The computational survey of electronic structure of metals by Moruzzi *et al.* [256] is a landmark example of the original KKR formalism, using the muffin-tin model and a root-search algorithm. The historical development of MST, from Huygens' principle and the original formulation by Lord Rayleigh [337, 338] to recent applications of full-potential theory, is surveyed by Gonis and Butler [148]. The full-potential theory was first proposed by Morgan and Williams [431, 432]. Nonspherical cell boundaries defined by the walls of polyhedral atomic cells were replaced by nonspherical enclosing potentials. This methodology was simplified when it was recognized [39, 40] that viable basis orbitals could be computed within the enclosing sphere S_1, and then matched after computation to a variational wave function at the polyhedral cell boundaries. Ultimately, it was shown that these two apparently disparate approaches necessarily produce the same orbital basis functions within the enclosed atomic cell [277]. In the surface integral formalism developed here, it is clear that the continuation of basis functions outside a local cell boundary cannot affect the C and S matrices that characterize MST. This formalism also provides a proof of the validity of MST for nonspherical atomic cells and potentials, which was disputed over a decade [148].

7.2 Variational principles

7.2.1 Kohn–Rostoker variational principle

Kohn and Rostoker [204] (KR) derived the KKR method using variational theory with a functional designed to produce the Lippmann–Schwinger equation. Local basis solutions can be computed to high accuracy within each unit cell. The variational formalism optimizes the construction of linear combinations of these local solutions that are compatible with the functions propagated from all adjacent cells. In this version of MST, local solutions are coupled through the free-scattering Green function of the Helmholtz equation. The Kohn–Rostoker variational functional [204],

$$\Lambda = \int_{\Re^3} \psi^* v \left(\psi - \int_{\Re^3} G_0 v \psi \right)$$

is stationary for infinitesimal variations of the one-electron wave function ψ if and only if it satisfies the homogeneous Lippmann–Schwinger equation, $\psi = \int_{\Re^3} G_0 v \psi$. The Green function G_0 satisfies the inhomogeneous Helmholtz equation at a specified energy $\epsilon = \kappa^2$ (Ryd),

$$(\nabla^2 + \kappa^2) G_0(\mathbf{x}, \mathbf{x}') = \delta(\mathbf{x}, \mathbf{x}').$$

The KR variational principle determines a wave function with correct boundary conditions at a specified energy, the typical conditions of scattering theory. Energy values are deduced from consistency conditions.

The KR variational principle and the resulting KKR/MST formalism determine wave functions for bound energy levels of a molecule or of a regular periodic solid or two-dimensional periodic surface. In these applications, the homogeneous Lippmann–Schwinger equation has no valid solutions except at a discrete set of energy levels. Hence either κ must be adjusted to meet the conditions for such a solution, or else the potential function must be altered to $v(\mathbf{x}) - \Delta\epsilon$ so that the parameter $\Delta\epsilon$ can be varied while retaining the same G_0. Using the positive parameter κ^2 to define G_0 already implies that energy values are measured relative to the minimum of an energy band. Because it simplifies functional forms of the Green function and solid harmonics, $\kappa^2 = 0$ is often used in LMTO calculations [384].

In a true scattering problem, an incident wave is specified, and scattered wave components of ψ are varied. In MST or KKR theory, the fixed term χ in the full Lippmann–Schwinger equation, $\psi = \chi + \int G_0\psi$, is required to vanish. χ is a solution of the Helmholtz equation. In each local atomic cell τ of a space-filling cellular model, any variation of ψ in the orbital Hilbert space induces an infinitesimal variation of the KR functional of the form $\delta\Lambda = \int_\tau \delta\psi^* v(\psi - \int_{\Re^3} G_0 v\psi) + hc$. This can be expressed in the form

$$\delta\Lambda = \int_\tau \delta\psi^* v(\chi^{in} - \chi^{out}) + hc,$$

where, integrating the finite volume integrals by parts, and using the coefficient matrices defined in MST,

$$\chi^{in} = \psi - \int_{\tau_\mu} G_0 v\psi = \sum_L J_L C_L^\mu,$$

$$\chi^{out} = \sum_{\nu\neq\mu} \int_{\tau_\nu} G_0 v\psi = -\sum_{\nu\neq\mu} \sum_{L,L'} J_L g_{L,L'}^{\mu\nu} S_{L'}^\nu.$$

In a finite basis of primitive functions ϕ_L^μ, the stationary condition is

$$-\delta\Lambda = \sum_{L,L'} \delta\gamma_L^{\mu\dagger} \sum_{L_1,L_2} S_{LL_1}^{\mu\dagger} \sum_\nu (\delta_{L_1L_2}^{\mu\nu} C_{L_2L'}^\nu + g_{L_1L_2}^{\mu\nu} S_{L_2L'}^\nu)\gamma_{L'}^\nu + hc = 0,$$

which implies the standard MST equations. Energy eigenvalues are determined by adjusting the trial energy so that the secular determinant of the matrix of MST equations is singular. Continuity conditions at cell boundaries are not addressed by this variational principle. It is usually assumed that ψ lies in the orbital Hilbert space (continuous function value and normal gradient at any cell interface). These conditions are required for consistency with the Schrödinger equation, but are not required by the KR variational principle and the MST equations.

7.2.2 Convergence of internal sums

In the Kohn–Rostoker variational equation, the summation indices refer to two different classes of functions: orbital basis functions ϕ_L, and solid harmonics J_L for the matrix of structure constants. For smooth matching at cell boundaries, the J_L sums should be carried to completion. In practice, finite sets of basis orbital functions and of solid harmonics are used. The set ϕ_L is truncated at some maximum orbital angular momentum $\bar{\ell}_\phi$ in each inequivalent atomic cell, and the local expansion of the Helmholtz Green function is limited to some maximum $\bar{\ell}_g$. In the muffin-tin model [204], a spherically averaged local potential does not couple different spherical harmonics in a single cell, but a propagated wave function in general has an infinite expansion in other cells because of the Green function. The **k**-dependent Green function for a periodic lattice structure introduces angular-momentum coupling into the translational cell even if it is spherical.

Model studies of full-potential KKR theory were carried out for both two-dimensional [51] and three-dimensional [52] space-lattices. The essential conclusion is that the truncation parameters $\bar{\ell}_\phi$ and $\bar{\ell}_g$ influence convergence in different ways, and should not ordinarily be set to a common value $\bar{\ell}$. Adequately converged results in general require essentially complete convergence of the geometrical expansion, which determines the Green function in the KR variational principle. In contrast, high angular momentum values are less significant for the shell structure of an atomic cell, considered to be a distorted atom. This can be understood because high orbital ℓ-values imply a centrifugal potential that excludes any orbital wave function if the classical turning radius is greater than the enclosing radius of the cell. Thus Rydberg orbitals in solids have no direct physical meaning if the Rydberg radius extends into the valence shell of adjacent atoms. For these reasons, the C and S matrices should be treated as rectangular column matrices, with the row index $\bar{\ell}_g$ significantly greater than the column index $\bar{\ell}_\phi$ [52].

Because rectangular matrices are singular, effective closure of these internal sums over solid-harmonic indices requires a generalized definition of $t = -SC^{-1}$ and of other expressions that involve inverse matrices in MST [280]. As the index L increases, the standard coefficient matrices S and C have very different behavior. The effect of the centrifugal barrier for high orbital ℓ-values is to make the regular solid harmonics J_L very small on any finite cell boundary, while standard normalization forces the irregular harmonics to be very large. In this Born-approximation limit, the low-order matrix C is extended by a unit matrix, while S is extended by a matrix of zeroes [281]. Thus the matrix $t = -SC^{-1}$ is augmented by a null matrix, while t^{-1} is augmented by elements of rapidly increasing magnitude, whose computational numerical errors can overwhelm the computationally accurate low-order elements. This can account for the strikingly different behavior found in empty-lattice calculations on the square-planar lattice using the KKR expression $t^{-1} - g$ [107] and using the formally equivalent expression $I - tg$ [438]. The latter calculations demonstrated exponential convergence with increasing L indices, while reproducing the much slower and less systematic convergence of the former calculations when the inverse matrix t^{-1} was explicitly truncated.

This problem can be avoided by expressing the MST equations in terms of the square matrix $S^\dagger C = C^\dagger S$, Hermitian in consequence of the surface-matching theorem. This matrix has full rank because it is contracted over the larger index ℓ_g. From the definitions of the C and S matrices, the matrix product $S^\dagger C$ is a specific integral involving the Helmholtz Green function [281],

$$\sum_{L''} S^{\mu\dagger}_{L,L''} C^{\mu}_{L'',L'} = -\int_{\tau_\mu} \phi^{\mu *}_L v \left(\phi^\mu_{L'} - \int_{\tau_\mu} G_0 v \phi^\mu_{L'} \right),$$

which can be evaluated in closed form. The contracted matrix product gS can also be evaluated in closed form by defining a generalized C-matrix, for displaced irregular solid harmonics N, $\tilde{C}^{\mu\nu}_{L_1,L} = (N^\mu_{L_1}|W_\nu|\phi^\nu_L)$. The function $N^\mu_{L_1}$ defined in cell τ_μ is evaluated on the surface σ_ν of a displaced cell. This is justified because N-functions are valid outside their cell of definition. Using the Green function to propagate N^μ, this becomes $\tilde{C}^{\mu\nu}_{L_1,L} = -\sum_{L_2} g^{\mu\nu}_{L_1 L_2} (J^\nu_{L_2}|W_\nu|\phi^\nu_L) = \sum_{L_2} g^{\mu\nu}_{L_1 L_2} S^\nu_{L_2,L}$. In a matrix notation omitting L indices and summations, the KKR/MST equation is

$$S^{\mu\dagger} \sum_\nu (\delta^{\mu\nu} C^\nu + g^{\mu\nu} S^\nu) \gamma^\nu = \sum_\nu \Lambda^{\mu\nu} \gamma^\nu = 0,$$

where all matrix products are contracted over the long dimension. In the Hermitian matrix Λ the triple product $S^{\mu\dagger} g^{\mu\nu} S^\nu = S^{\mu\dagger} \tilde{C}^{\mu\nu} = (\tilde{C}^\dagger)^{\mu\nu} S^\nu$. Multiplying on the left by $S(S^\dagger C)^{-1}$, and using $(\tilde{C}^\dagger)^{\mu\nu} = S^{\mu\dagger} g^{\mu\nu}$, this takes the form $\Xi\beta = 0$, defining a modified secular matrix

$$\Xi^{\mu\nu} = \delta^{\mu\nu} + S^\mu (S^{\mu\dagger} C^\mu)^{-1} (\tilde{C}^\dagger)^{\mu\nu}.$$

The coefficients $\beta = S\gamma$ are the expansion coefficients of the global matching function $\xi(\sigma) = \sum_{\mu,L} N^\mu_L(\mathbf{r}_\sigma)\beta^\mu_L$. Defining $t = -S(S^\dagger C)^{-1} S^\dagger$, which reduces to $t = -SC^{-1}$ for square matrices [281], $\Xi^{\mu\nu}$ takes the standard MST form $I - tg$, without implying inversion of a rectangular matrix. Using these closed forms, the coefficients β and γ are related by

$$\gamma^\mu = -(S^{\mu\dagger} C^\mu)^{-1} \sum_{\nu \neq \mu} (\tilde{C}^\dagger)^{\mu\nu} \beta^\nu.$$

In alloy theory, this equation determines the variational wave function for an atomic cell τ_μ embedded in a statistical medium defined by the vector of coefficients β for all other cells [157, 281].

7.2.3 Schlosser–Marcus variational principle

Since a large truncation parameter $\bar{\ell}_g$ implies large unsymmetrical C and S matrices, and a very large matrix of structure constants, it might be more efficient to find a variational condition directly applicable to the matching conditions for wave functions on the common interface between adjacent atomic cells. Schlosser and Marcus [359] (SM) extended the Schrödinger variational principle to the case of basis functions that might be discontinuous or have discontinuous normal gradients across atomic cell boundaries. It will be shown here that this variational principle is equivalent to that of Kohn and Rostoker if the internal sums over solid harmonic L-indices are carried to completion in the MST formalism. The SM variational principle has been used by Ferreira and Leite [112, 113, 110] to formulate the variational cellular method (VCM), which eliminates the need for structure constants. A matrix of Wronskian surface integrals over all shared cell facets is constructed. Optimal matching conditions correspond to zeroes of the determinant of this matrix.

The Schlosser–Marcus variational principle is derived for a single surface σ that subdivides coordinate space $\Re^3$ into two subvolumes τ_{in} and τ_{out}. This generalizes immediately to a model of space-filling atomic cells, enclosed for a molecule by an external cell extending to infinity. The continuity conditions for the orbital Hilbert space require $\psi^{out} \equiv_\sigma \psi^{in}$. This implies a vanishing Wronskian surface integral

$$(\psi^{in}|W_\sigma|\psi^{out}) = -\int_\sigma [(\psi^{in})^* \nabla_n \psi^{out} + (\nabla_n \psi^{in})^* \psi^{out}] = 0,$$

where σ is the enclosing boundary of τ_{in}. The sign of the first term here is reversed because the notation W_σ implies an outward normal derivative $in \rightarrow out$, while by convention, $\nabla_n \psi$ denotes the outward normal gradient across the boundary that encloses the region of definition of ψ. The notation W_μ will be used for a Wronskian integral with outward normal gradients over the full bounding surface of atomic cell τ_μ. The SM variational functional adds a Wronskian surface integral $Z_\sigma = \frac{1}{2}(\psi^{in} + \psi^{out}|W_\sigma|\psi^{in} - \psi^{out})$ to the Schrödinger functional $Z_\tau = \int_{\Re^3} \psi^*(\mathcal{H} - \epsilon)\psi$. This interface surface integral vanishes if $\psi^{in} \equiv_\sigma \psi^{out}$ for independent trial functions ψ^{in} and ψ^{out}. Then the global trial function ψ is in the usual Hilbert space and the variational condition $\delta Z = 0$ is just the Schrödinger condition for stationary energy ϵ.

Schlosser and Marcus [359] showed that for variations about such a continuous trial function, the induced first-order variations of Z_τ and Z_σ exactly cancel, even if the orbital variations are discontinuous at σ or have discontinuous normal gradients. After integration by parts, the variation of Z_τ about an exact solution is a surface Wronskian integral

$$\delta Z_\tau = -(\psi|W_\sigma|\delta\psi^{in} - \delta\psi^{out}).$$

The variation of Z_σ is

$$\delta Z_\sigma = (\psi|W_\sigma|\delta\psi^{in} - \delta\psi^{out}),$$

which exactly cancels δZ_τ [359].

Because of this cancellation, variation about an arbitrary trial function gives

$$\delta(Z_\tau + Z_\sigma) = (\delta\psi|\mathcal{H} - \epsilon|\psi) + \frac{1}{2}(\delta\psi^{in} + \delta\psi^{out}|W_\sigma|\psi^{in} - \psi^{out}).$$

This can vanish only if $(\mathcal{H} - \epsilon)\psi = 0$ in both τ_{in} and τ_{out}. Moreover, this requires that both $(\delta\psi^{in}|W_\sigma|\psi^{in} - \psi^{out})$ and $(\delta\psi^{out}|W_\sigma|\psi^{in} - \psi^{out})$ must vanish when ψ^{in} and ψ^{out} are varied independently. By an extension of the surface matching theorem, both these Wronskian integrals must vanish in order to eliminate the value and normal gradient of $\psi^{in} - \psi^{out}$ on σ. Practical applications of this formalism use independent truncated orbital basis expansions in adjacent atomic cells, so that the continuity conditions cannot generally be satisfied exactly.

In a space-filling cellular model, the SM variational functional can be expanded in a local basis in each atomic cell. Variation of the expansion coefficients of the trial orbital function $\psi = \sum_L \phi_L^\mu \gamma_L^\mu$ in cell τ_μ induces the variation

$$\delta Z = \sum_{L,L'} \delta\gamma_L^{\mu\dagger} \sum_\nu Z_{LL'}^{\mu\nu} \gamma_{L'}^\nu + hc.$$

Matrix elements $Z_{LL'}^{\mu\nu}$ are Wronskian surface integrals on interface $\sigma_{\mu\nu}$ between adjacent cells τ_μ and τ_ν, evaluated for basis functions $\phi_L^\mu, \phi_{L'}^\nu$. If the basis functions in cell μ are all evaluated by integrating the local Schrödinger equation at the same energy, the site-diagonal matrix elements $Z^{\mu\mu}$ vanish. The interface elements are

$$Z^{\mu\nu} = -\frac{1}{2}(\phi^\mu|W_{\mu\nu}|\phi^\nu), \qquad Z^{\nu\mu} = Z^{\mu\nu\dagger},$$

where $W_{\mu\nu}$ implies normal gradients in the sense $\mu \to \nu$ on the cell interface $\sigma_{\mu\nu}$. By this convention, matrix $Z^{\mu\nu}$ is Hermitian.

The variational cellular method (VCM) [113, 110] is an application of the Schlosser–Marcus variational principle. The VCM variational matrix omits cell-diagonal Wronskian integrals, which vanish by construction in a basis of orbital functions all computed at the same energy. Because this matrix is Hermitian, its null eigenvalues define solutions of the variational problem [113]. The SM surface functional vanishes not only if ψ^{out} and ψ^{in} match in both value and normal gradient at σ, but also if they are both equal and opposite. This implies that the VCM determinant vanishes both for "true" solutions, giving an optimal solution of the surface matching conditions, and for "false" solutions, giving a maximal mismatch. This can be understood in a one-dimensional model, in which the Wronskian condition matches logarithmic derivatives across a boundary. The correct logarithmic derivative is obtained even if both function value and gradient have the wrong sign. The VCM formalism avoids such false solutions by retaining only those zeroes of the secular determinant for which the energy derivative is negative, a criterion that selects the "true" solutions [113].

If cell-diagonal terms are retained, the VCM variational equations are

$$\sum_{\nu\neq\mu} (\phi_L^\mu | W_{\mu\nu} | \psi^\mu - \psi^\nu) = 0,$$

for each atomic cell surface σ_μ, summed over all adjacent cells. Although two surface integral conditions are needed in general to eliminate both value and gradient discontinuities, these equations can be justified because each interface $\sigma_{\mu\nu}$ occurs twice, for cells μ and ν, respectively. Expansion in a local orbital basis in each cell gives

$$\sum_{\nu\neq\mu} \left(\phi_L^\mu | W_{\mu\nu} | \sum_{L'} \phi_{L'}^\mu \gamma_{L'}^\mu - \sum_{L'} \phi_{L'}^\nu \gamma_{L'}^\nu \right) = 0.$$

For an exact solution, the external function ψ^ν would be identical to the global matching function ξ on each interface. An alternative algorithm can be based on fitting a linear combination of basis functions in each cell to ξ, which is uniquely determined. The VCM equations for an orbital basis at fixed energy reduce to [282]

$$\left(\phi_L^\mu | W_\mu | \sum_{L'} \phi_{L'}^\mu \gamma_{L'}^\mu - \xi \right) = -\,(\phi_L^\mu | W_\mu | \xi) = 0.$$

The global matching function determined by MST is

$$\xi = \sum_{\mu,L} N_L^\mu \beta_L^\mu = \sum_L \left(N_L^\mu \beta_L^\mu - J_L^\mu \sum_{\nu\neq\mu} \sum_{L'} g_{L,L'}^{\mu\nu} \beta_{L'}^\nu \right),$$

expanded using coefficients

$$\beta_L^\nu = -(J_L^\nu | W_\nu | \xi) = -\sum_{L'} (J_L^\nu | W_\nu | \phi_{L'}^\nu) \gamma_{L'}^\nu = \sum_{L'} S_{L,L'}^\nu \gamma_{L'}^\nu.$$

The variation of the SM functional in τ_μ then reduces to

$$2\delta Z = \delta\gamma^{\mu\dagger} (\phi^\mu | W_\mu | J^\mu) \left[(N^\mu | W_\sigma | \phi^\mu) \gamma^\mu - \sum_{\nu\neq\mu} g^{\mu\nu} (J^\nu | W_\nu | \phi^\nu) \gamma^\nu \right] + hc,$$

omitting L-indices and summations to simplify the notation. The first term is derived using the general formula

$$\sum_L (\phi_i|W_\sigma|J_L)(N_L|W_\sigma|\phi_j) = \sum_L (\phi_i|W_\sigma|N_L)(J_L|W_\sigma|\phi_j),$$

from the surface matching theorem. $2\delta Z$ agrees with the KR variational expression if intermediate sums over angular indices are complete [148]. Because the SM variational principle implies stationary energy eigenvalues when the continuity conditions are exactly satisfied, this is also true for the KKR/MST equations when variational trial functions are continuous with continuous gradients, as for muffin-tin and atomic-cell orbitals.

7.2.4 Elimination of false solutions

In a root-search procedure, false zeroes of the VCM secular determinant can easily be identified by testing its energy derivative [113]. However, this is a major obstacle to energy linearization of the method, following the logic of the LMTO method as a linearization of KKR [384]. Two simplified variants of VCM have been proposed that eliminate "false" solutions at the cost of abandoning a variational formalism related to Schrödinger's condition of stationary energy [287]. The equations of the Green-function cellular method (GFCM) [53, 444, 282], or NVCM [287], are

$$\sum_{\nu\neq\mu}\left(N^\mu_L|W_{\mu\nu}|\sum_{L'}\phi^\mu_{L'}\gamma^\mu_{L'} - \sum_{L'}\phi^\nu_{L'}\gamma^\nu_{L'}\right) = 0. \tag{7.6}$$

An alternative formalism (JVCM [287]) substitutes the equations [443]

$$\sum_{\nu\neq\mu}\left(J^\mu_L|W_{\mu\nu}|\sum_{L'}\phi^\mu_{L'}\gamma^\mu_{L'} - \sum_{L'}\phi^\nu_{L'}\gamma^\nu_{L'}\right) = 0. \tag{7.7}$$

A variational alternative, not yet implemented, is to augment the SM variational functional by a Lagrange-multiplier term designed to enforce the matching condition at each atomic cell boundary. The proposed variational functional Ξ supplements the Schrödinger functional $\Xi_\tau = \int_{\Re^3}\psi^*(\mathcal{H}-\epsilon)\psi$ by a Wronskian integral over the global matching surface,

$$\Xi_\sigma = \left(\frac{1}{2}\psi^{in} + \frac{1}{2}\psi^{out} - \xi|W_\sigma|\psi^{in} - \psi^{out}\right).$$

ξ and its normal gradient, defined on the global matching surface, are both Lagrange-multiplier fields, to be determined such that $\psi^{in} \equiv_\sigma \psi^{out}$. Comparison with MST shows that if this condition is satisfied, ξ is the global matching function

$\xi \equiv_\sigma \psi^{out} \equiv_\sigma \psi^{in}$. Then Ξ_σ vanishes quadratically and Ξ reduces to the Schrödinger functional for trial functions ψ in the usual Hilbert space.

If Ξ_σ is stationary with respect to free variations of ψ^{in}, ψ^{out}, and ξ,

$$\delta\Xi_\sigma = \left(\frac{1}{2}\psi^{in} + \frac{1}{2}\psi^{out} - \xi|W_\sigma|\delta\psi^{in} - \delta\psi^{out}\right) + \left(\frac{1}{2}\delta\psi^{in} + \frac{1}{2}\delta\psi^{out} - \delta\xi|W_\sigma|\psi^{in} - \psi^{out}\right) = 0.$$

This implies

$$\xi \equiv_\sigma \frac{1}{2}(\psi^{in} + \psi^{out}); \qquad \psi^{in} \equiv_\sigma \psi^{out},$$

or $\psi^{in} \equiv_\sigma \psi^{out} \equiv_\sigma \xi$. Hence ξ is the global matching function, and a global trial function ψ is defined in the usual Hilbert space. Variation of this function does not produce any surface terms in $\delta\Xi$. The implied practical procedure, as in the LMTO method [384], is to use these variational conditions to construct trial functions as a basis for the Schrödinger eigenvalue equations.

The variational equations imply $\psi - \xi \equiv_\sigma 0$ on each cell boundary σ_μ. Given independent expansions $\psi^\mu = \sum_L \phi_L^\mu \gamma_L^\mu$ within each atomic cell, and $\xi(\sigma) = \sum_{\mu,L} N_L^\mu(\mathbf{r}_\sigma)\beta_L^\mu$ on the global matching surface, the coefficients are determined by the implied variational equations. The surface matching theorem implies two independent Wronskian integral conditions for each atomic cell,

$$\left(J_L^\mu|W_\mu|\psi^\mu - \xi\right) = -\sum_{L'} S_{LL'}^\mu \gamma_{L'}^\mu + \beta_L^\mu = 0,$$

$$\left(N_L^\mu|W_\mu|\psi^\mu - \xi\right) = \sum_{L'} C_{LL'}^\mu \gamma_{L'}^\mu - \sum_{L_1,L'}\sum_{\nu\neq\mu} \left(N_L^\mu|W_\mu|N_{L_1}^\nu\right) S_{L_1L'}^\nu \gamma_{L'}^\nu = 0.$$

The first condition determines the expansion coefficients $\beta^\mu = S^\mu\gamma^\mu$. On substituting $(N_L^\mu|W_\mu|N_{L_1}^\nu) = -\sum_{L_2}(N_L^\mu|W_\mu|J_{L_2}^\mu)g_{L_2L_1}^{\mu\nu} = -g_{LL_1}^{\mu\nu}$, the second condition reduces to the standard MST equations,

$$\sum_{L'} C_{LL'}^\mu \gamma_{L'}^\mu + \sum_{\nu\neq\mu}\sum_{L_1,L'} g_{LL_1}^{\mu\nu} S_{L_1L'}^\nu \gamma_{L'}^\nu = 0.$$

An attempt to fit ξ on the boundary of any cell τ_μ by a single expansion in the local primitive basis functions ϕ_L fails if these functions all satisfy the same Schrödinger equation at a single energy. The problem is that such basis functions all have vanishing Wronskian surface integrals with each other, and cannot match both value and normal gradient of an arbitrary function on the surface σ_μ. In MST the set of irregular solid harmonics N^μ from all cells is not limited by this constraint, and thus provides a suitable basis for ξ. In contrast, representation as a sum of local primitive basis orbitals in a single cell leads to the problem of "false" solutions of the SM variational equations. The MST expansion of ξ is suggestive that, on any interface segment $\sigma_{\mu\nu}$, ξ must be represented by two distinct terms, one from the indexed cell μ and the other from cell ν. In general this ensures that the number of matching conditions (value and normal gradient for each basis function) equals the number of independent basis functions.

7.3 Energy-linearized methods

Because of the nonlinear energy dependence of the KKR secular equation, roots of the secular determinant must be computed by a root-search procedure, requiring repeated evaluation of the secular matrix for each root. Despite the inherent inefficiency of this procedure, simplifications implicit in the KKR–DFT theory for muffin-tin potential functions can be exploited. This methodology was used [256] to compute self-consistent energy-band structures for 32 elemental metals, ranging from H to In in the periodic table, including spin-polarized structures for the transition metals. This reference provides a bibliography giving technical details of the analytical and computational methods used. In order to carry out self-consistent calculations on more complex materials with polyatomic translational cells, or to examine effects of lattice distortion, it is desirable to convert this formalism into a variational form in which energy eigenvalues can be computed from a linearized secular equation.

7.3.1 The LMTO method

Starting from the idea of "muffin-tin orbitals" (MTO) [7], the linear muffin-tin orbital method (LMTO), developed by Andersen and collaborators, became one of the most widely used computational methods in energy-band theory. Publications on this method include a general description, with source listings of computer programs [384], and a review of methodology and applications [11]. The LMTO method differs from the KKR muffin-tin model of multiple scattering theory in three important aspects. In addition to conversion of the theory to an energy-linearized form, the method also replaces the original muffin-tin model by an atomic-sphere approximation (ASA), in which the local atomic potential function is extended from the enclosed muffin-tin sphere S_0 to a larger "atomic" sphere, defined to have the same volume as the Wigner–Seitz atomic cell. The third element of difference from KKR is that structure constants are taken to be independent of energy. This is done by subtracting an energy shift from both sides of the Schrödinger equation,

$$\left(\nabla^2 + \kappa_0^2\right)\psi(\mathbf{x}) = \left(v(\mathbf{x}) - \kappa^2 + \kappa_0^2\right)\psi(\mathbf{x}). \tag{7.8}$$

The Green function G_0 and solid-harmonic functions J_L and N_L are required only for the fixed energy κ_0^2, usually taken to be zero. This amounts to adding a linear energy term to the potential function in the interstitial region, justified in full-potential theory or the ASA because the net interstitial volume is reduced to zero. The practical effect is slower convergence of the angular momentum expansions. Although the ASA violates the original KKR conditions that ensure an exact solution of the muffin-tin model problem, the ASA model of overlapping atomic spheres is in many ways a more satisfactory approximation to full-potential MST. In particular, in the surface integral formalism, approximating Wronskian integrals over the surface of a Wigner–Seitz cell by the corresponding integrals over an equivalent-volume sphere can be considered as a first approximation to a surface quadrature scheme for polyhedral cells. The full DFT local potential function can be used within the atomic sphere. The muffin-tin model, which assumes a constant potential in the near-field region, has no special justification as a computational approximation for atomic potentials.

In the usual ASA, the t-matrix is real and diagonal, but the elements have both zeroes and poles as functions of energy. The original proposal was to simplify KKR calculations by using constant-energy structure constants while fitting diagonal elements of t^{-1} in the standard secular equation, $(t^{-1} - g)\beta = 0$, to rational functions of energy [8]. This also simplifies the energy dependence of muffin-tin orbitals, which can be used as basis functions in the standard Schrödinger variational principle for the energy eigenvalues. The resulting LMTO formalism (called LCMTO in the original presentation) produces a linear eigenvalue problem that includes a "combined correction" matrix representing the difference of the overlap matrix evaluated in atomic spheres versus atomic polyhedra together with a correction for basis functions of higher ℓ quantum numbers. LMTO energy-band calculations obtain all energy eigenvalues at a given $\mathbf{k}$-point from a single matrix diagonalization. These eigenvalues are valid within an energy panel in which the rational fit to the t^{-1}-matrix is sufficiently accurate.

In considering nonspherical local potentials, a simple rational fit to diagonal elements of t^{-1} is no longer valid. From the general definition of the t-matrix as $-SC^{-1}$ in terms of the standard MST matrices, it is clear that for energy-independent solid-harmonic functions, linear energy expansion of the local basis functions $\phi_L(\epsilon; \mathbf{r})$ is equivalent to fitting elements of the t-matrix by a simple rational formula. In the ASA this produces the parametrization proposed by Andersen. It was recognized at the same time [13] that a linear energy expansion of the basis functions also solves the practical problem in MTO theory that in the local representation $\chi = \phi - \sum JC$ of primitive MTOs the regular solid-harmonic functions are not orthogonal to inner-shell occupied orbitals. The practical solution is to replace the solid harmonics by linear combinations of the energy-derivative functions denoted here by $\dot{\phi}_L$. In the standard LMTO method, the basis functions are energy-independent MTOs expressed as linear combinations of functions ϕ_L and $\dot{\phi}_L$ computed at a fixed panel energy [9, 384]. Basis functions of this kind are used in alternative energy-linearized methods [9, 430] which will not be considered here.

7.3.2 The LACO method

The linearized atomic-cell orbital (LACO) method is a full-potential analog of LMTO [274]. Energy-independent structure constants are computed using the LMTO program STR [384]. On a grid of energy values suitable for accurate interpolation, basis functions are computed by numerical integration within the enclosing sphere of each atomic cell. The basis functions and their normal gradients are interpolated to atomic-cell boundaries so that the standard MST C and S matrices can be evaluated by two-dimensional quadrature over the surfaces of polyhedral atomic Wigner–Seitz cells. Energy linearization is accomplished in two stages, each with several options as to the specific procedure. In the first stage, energy-independent atomic-cell orbitals are constructed in the form $\phi + \sum \dot{\phi}\omega$. Here, ϕ is a single primitive basis function ϕ_L^μ in a particular cell τ_μ. In this cell, the defining ACO form $\phi - \sum JC$ is modified by representing the regular solid-harmonic functions J as linear combinations of energy-derivative functions $\dot{\phi}_L$. In all other cells, the matched external function $\sum NS$ is expanded in the local $\dot{\phi}_L$ basis. The coefficient matrix $\omega_{L,L'}$ satisfies a set of inhomogeneous linear equations, obtained from one of the linear forms of full-potential MST equations, in the extended fixed-energy basis of functions ϕ_L and $\dot{\phi}_L$. In the second stage of LACO calculations, these energy-independent ACO functions are used in the standard Rayleigh–Schrödinger or Schlosser–Marcus variational equations to determine energy eigenvalues.

In an energy range or panel sufficiently narrow that linear energy interpolation of the basis functions is accurate, it can be shown that this general linearization procedure produces eigenfunctions identical to the null vectors of the MST secular equations. To prove this for basis functions all at the same energy ϵ_0, consider right-eigenvectors of the complex, unsymmetric matrix ω. The eigenvalue equation $\omega c_k = c_k \Delta\epsilon_k$ defines a displacement $\Delta\epsilon$ from the fixed panel energy. Each eigenvector of the ω-matrix determines a linear combination of energy-independent ACO functions which, by construction, is an optimized fit to the orbital continuity conditions. This function is

$$\psi_k = \sum \left(\phi + \sum \dot{\phi}\omega\right) c_k = \sum (\phi + \dot{\phi}\Delta\epsilon_k) c_k = \sum \phi(\epsilon_k) c_k. \quad (7.9)$$

This agrees with the form of an MST null-vector expanded in the energy-dependent basis at $\epsilon_k = \epsilon_0 + \Delta\epsilon_k$. Because of numerical approximations inherent in practical calculations, it is generally inconvenient to diagonalize matrix ω directly. The eigenvalues are complex numbers if there are any residual numerical errors. For this reason, a standard variational principle based on Hermitian matrices is used in the second stage of linearized calculations. The $\phi, \dot{\phi}$ expansion must be used with caution unless all elements of the ω-matrix are small. A large energy shift $\Delta\epsilon$ may imply that basis functions fall outside the energy panel used in a particular calculation, so that the interpolated functions are not accurate. So-called "ghost" bands can occur, especially for higher angular quantum numbers, when functions ϕ and $\dot{\phi}$ tend to become linearly dependent within a local cell [11].

7.3.3 Variational theory of linearized methods

The Kohn–Rostoker variational principle [204] implies variational equations in the extended basis $\{\phi, \dot{\phi}\}$. Each energy-independent ACO basis function $\tilde{\phi}^\lambda$ is defined by one fixed function ϕ^λ modified by a sum of $\dot{\phi}^\mu$ functions with coefficients $\omega^{\mu\lambda}$. Suppressing L-indices, the KKR/MST equations indexed by $\tilde{\phi}^\kappa$ in cell τ_μ for the coefficients in $\tilde{\phi}^\lambda$ are

$$\sum_\nu (S^\mu \delta^{\mu\kappa} + \dot{S}^\mu \omega^{\mu\kappa})^\dagger [\delta^{\mu\nu}(C^\nu \delta^{\nu\lambda} + \dot{C}^\nu \omega^{\nu\lambda}) + g^{\mu\nu}(S^\nu \delta^{\nu\lambda} + \dot{S}^\nu \omega^{\nu\lambda})] = 0. \quad (7.10)$$

Matrices $C, S, \dot{C}$ and $\dot{S}$ here are to be considered as rectangular matrices. The internal sums over solid-harmonic L-indices should be carried to convergence. The L', L indices of the square matrix ω are basis function indices and may have a smaller range.

If the KKR functional Λ were treated as a functional of the coefficient matrix ω, the derived variational equations would be a set of linear equations of the form $\dot{S}^{\mu\dagger} \sum_\nu [\cdots] = 0$, where the bracketed term is the same as in Eqs. (7.10). The solution of these simplified equations for a given value of λ, L and all values of μ, L' is a column vector of the ω-matrix. These simplified equations were tested by empty-lattice calculations on an fcc space-lattice [280].

The Schlosser–Marcus variational principle [359] provides an alternative that does not use structure constants. On substituting the expansion of an energy-independent ACO into the SM variational functional, the variational equations indexed by $\tilde{\phi}^\kappa$ in cell τ_μ are

$$\sum_{\nu \neq \mu} (\phi^\mu \delta^{\mu\kappa} + \dot{\phi}^\mu \omega^{\mu\kappa}) | W_{\mu\nu} | (\phi^\mu \delta^{\mu\lambda} + \dot{\phi}^\mu \omega^{\mu\lambda}) - (\phi^\nu \delta^{\nu\lambda} + \dot{\phi}^\nu \omega^{\nu\lambda}) = 0, \quad (7.11)$$

where L-indices and sums are suppressed to simplify the notation. In these equations, the internal summations in Eqs. (7.10) are replaced by direct matching across adjacent cell interfaces. If the generally complex unsymmetric matrix ω is diagonalized, with eigenvector elements c^μ_{Lk}, Eqs. (7.11) reduce to the VCM equations for $\psi^\mu = \sum_L \phi^\mu_L(\epsilon_k)c^\mu_{Lk}$,

$$\sum_{\nu\neq\mu}(\phi^\mu_L(\epsilon_k)|W_{\mu\nu}|\psi^\mu - \psi^\nu) = 0,$$

whose null vectors determine equivalent linear combinations of ACO functions or of atomic basis functions at energy values displaced by $\Delta\epsilon_k$. In this representation, the LACO method is an energy-linearized version of the variational cellular method (VCM) [112]. However, Eqs. (7.11) are quadratic in the matrix ω, which is nonhermitian in general and complex for periodic lattice structures, due to translational phase factors. The problem of "false" solutions, inherent in the SM/VCM variational principle, remains in these quadratic matrix equations, for which no computationally viable solution algorithm has been found. Approximate linear equations are obtained if the left-hand element is simplified to either just ϕ^μ or $\dot{\phi}^\mu$, as done in tests [280] of the corresponding KKR/MST equations. As in the case of Eqs. (7.10), each solution of the simplified equations for given λ, L is a column vector of the ω-matrix. The two alternative sets of simplified equations cannot be satisfied simultaneously.

These difficulties with the linearized VCM can be resolved by using the extended functional, defined over the global matching surface σ in Subsection 7.2.4, above,

$$\Xi_\sigma = \left(\frac{1}{2}\psi^{in} + \frac{1}{2}\psi^{out} - \xi|W_\sigma|\psi^{in} - \psi^{out}\right).$$

Following LMTO theory, an energy-independent ACO function $\tilde{\phi}^\lambda$ based on atomic cell τ_λ is represented in all cells τ_μ by the form $\phi^\mu\delta^{\mu\lambda} + \dot{\phi}^\mu\omega^{\mu\lambda}$. The Lagrange-multiplier field ξ, which becomes the global matching function in the variational equations, must be expanded in a basis of functions that can be extended to completeness on the global matching surface. In MST, the expansion on each interface sector $\sigma_{\mu\nu}$ of the matching surface consists of two terms. One is the local sum, $\sum_L N^\mu_L$, and the other is the corresponding sum over irregular solid harmonics extended to $\sigma_{\mu\nu}$ from all other cells. Both sums are needed in order to match both value and normal gradient of the wave function on the local cell surface. Because the primitive basis functions ϕ^μ_L are strictly truncated on σ_μ, the surface of cell τ_μ, the energy-derivative functions $\dot{\phi}^\mu$ should be defined for consistency as identically zero outside σ_μ. By implication, replacing solid-harmonic functions by the local basis $\{\dot{\phi}\}$ in each cell, the matching function should be expanded on each $\sigma_{\mu\nu}$ as $\xi = \dot{\phi}^\mu\beta^{\mu\lambda} + \dot{\phi}^\nu\beta^{\nu\lambda}$. The Wronskian integrals of $\psi^\mu - \xi$ that must vanish for each cell τ_μ are

$$\begin{aligned}\sum_{\nu\neq\mu}(\phi^\mu|W_{\mu\nu}|(\phi^\mu\delta^{\mu\lambda} + \dot{\phi}^\mu\omega^{\mu\lambda}) - (\dot{\phi}^\mu\beta^{\mu\lambda} + \dot{\phi}^\nu\beta^{\nu\lambda})) = 0,\\ \sum_{\nu\neq\mu}(\dot{\phi}^\mu|W_{\mu\nu}|(\phi^\mu\delta^{\mu\lambda} + \dot{\phi}^\mu\omega^{\mu\lambda}) - (\dot{\phi}^\mu\beta^{\mu\lambda} + \dot{\phi}^\nu\beta^{\nu\lambda})) = 0.\end{aligned} \tag{7.12}$$

If the total number of basis functions ϕ_L^μ for all cells is N_ϕ, then for each global solution index λ there are $2N_\phi$ equations for the $2N_\phi$ elements of the column vectors $\omega^{\mu\lambda}$ and $\beta^{\mu\lambda}$. Thus the variational equations derived from Ξ_σ provide exactly the number of inhomogeneous linear equations needed to determine the two coefficient matrices, ω and β. These equations have not yet been implemented, but they promise to provide an internally consistent energy-linearized full-potential MST.

7.4 The Poisson equation

For a solid or molecule subdivided into atomic cells, the Poisson equation, given in Rydberg units by

$$\nabla^2 v = -8\pi\rho, \tag{7.13}$$

can be solved by full-potential multiple scattering theory methods. Here ρ is the local number density of electrons, and the negative electronic charge is factored out of both charge density and potential. Adopting the classical theory of inhomogeneous linear differential equations to a cellular model, particular solutions are first obtained for each atomic cell and then modified by the addition of regular solutions of the homogeneous equation to satisfy continuity and boundary conditions. The original full-potential MST [432] was modified for the Poisson equation and applied to a periodic charge distribution on an fcc space-lattice [247]. In this example all atomic cells are geometrically equivalent, so the actual calculation is for a single cell with periodic boundary conditions. The given charge distribution, with Heaviside cutoff factors at the polyhedral cell boundary, was expanded in spherical harmonics at radii up to that of the enclosing sphere. A local particular solution v_0 was obtained by integrating equations analogous to the KKR/MST equations in spherical polar coordinates, subject to the boundary condition that the particular solution should vanish at large r. The external continuation of v_0 is a sum of multipole potentials. Within a particular cell, the sum of these potentials due to all other cells defines a local potential Δv, expressed in terms of structure constants that are just the coefficients in the two-center expansion of the Coulomb interaction. Δv is added to v_0 to give the required global solution of the Poisson equation. The surface integral MST method described below has been applied to the Poisson equation, and is incorporated in the LACO program package [278], still using structure constants. More recently, MST cellular methods, using direct matching theory with NVCM (GFCM) and JVCM equations rather than structure constants, were applied to this problem in a detailed numerical study [443]. This study finds that the JVCM is the most accurate and reliable of the methods considered, which included standard MST with structure constants.

Multiple scattering theory treats the Poisson equation as an inhomogeneous Helmholtz equation of zero energy (Laplace equation). The Green function G_0 is proportional to the Coulomb potential. In Rydberg units, $8\pi G_0 = -2/r_{12}$. A particular solution v_0 and complementary solution Δv within a local atomic cell τ are defined by subdividing the Lippmann–Schwinger equation such that

$$v = v_0 + \Delta v = -8\pi \int_\tau G_0 \rho - 8\pi \int_{\mathcal{R}^3 - \tau} G_0 \rho. \tag{7.14}$$

The choice of solid-harmonic functions

$$J_L = r^\ell Y_L/(2\ell + 1); \qquad N_L = r^{-\ell-1} Y_L$$

gives the well-known one-center expansion of $1/r_{12}$ [188]. It is assumed that the local density function is subdivided into spherical harmonic components as $\rho = \sum_L \rho_L$. Each of these components defines a particular solution within the enclosing sphere of the local cell. A very efficient numerical algorithm is available for solving the radial Poisson equation [229], generalized to arbitrary ℓ-values [278]. Denoting these primitive solutions by $\hat{v}_L$, the ACO construction can be used to define a particular solution v_{0L} for each ρ_L such that $v_{0L} = \hat{v}_L - \sum J\hat{c}$ in the local cell τ. v_{0L} matches onto an external function $\sum N\hat{s}$ on the cell surface σ. The coefficient matrices here are defined by

$$\hat{c}_{L',L} = (N_{L'}|W_\sigma|\hat{v}_L), \qquad \hat{s}_{L',L} = -(J_{L'}|W_\sigma|\hat{v}_L). \tag{7.15}$$

The complementary potential Δv or generalized Madelung term is expanded in regular solid harmonics as

$$\Delta v_L^\mu = -\sum_{L'} J_{L'}^\mu \Delta c_{L',L}^\mu. \tag{7.16}$$

The coefficients here are

$$\Delta c^\mu = -\sum_\nu g^{\mu\nu}(J^\nu|W_\nu|v_0^\nu), \tag{7.17}$$

obtained by substituting the two-center expansion of N^μ into the second term in Eq. (7.14), or, equivalently, into $\sum N\hat{s}$.

Alternative equations for the coefficients Δc are given by direct matching at cellular interfaces, bypassing the need for structure constants. The two alternatives considered by [443] are the NVCM or GFCM equations,

$$\sum_{\nu\neq\mu}(N^\mu|W_{\mu\nu}|J^\nu)\Delta c^\nu - \Delta c^\mu = \sum_{\nu\neq\mu}(N^\mu|W_{\mu\nu}|v_0^\nu), \tag{7.18}$$

and the JVCM equations,

$$\sum_{\nu\neq\mu}(J^{\mu}|W_{\mu\nu}|J^{\nu})\Delta c^{\nu} = \sum_{\nu\neq\mu}(J^{\mu}|W_{\mu\nu}|v_0^{\nu}) - (J^{\mu}|W_{\mu}|v_0^{\mu}). \tag{7.19}$$

Exploiting the high efficiency of the integration algorithm for the radial Poisson equation [229], MST surface integral formalism [274, 278] reduces the computation of multipole moments of cellular charge densities to the evaluation of Wronskian surface integrals. Electrostatic multipole moments Q_L are defined by the asymptotic external potential for each cell τ. For the particular choice of solid-harmonic functions given above,

$$\lim_{r\to\infty} v_{0L} = \sum N_{L'}\hat{s}_{L',L} = -8\pi\sum_L N_L Q_L. \tag{7.20}$$

A factor -2 included in the last term here compensates for the use of Rydberg units and for the omission of the negative electronic charge in potential functions derived from Eq. (7.14). Hence the electrostatic multipole moments of atomic cell τ_μ are

$$Q_L^{\mu} = -\frac{1}{8\pi}\sum_{L'}\hat{s}^{\mu}_{L,L'} = \frac{1}{8\pi}\left(J_L^{\mu}|W_{\sigma_\mu}|\sum_{L'}\hat{v}^{\mu}_{L'}\right). \tag{7.21}$$

This equation is used for $\ell = 0$ to evaluate normalization integrals in the LACO program package [278], avoiding numerical volume quadrature.

7.5 Green functions

Although electronic structure calculations are usually concerned with finding the eigenfunctions $\psi_i(\mathbf{x})$ and eigenvalues ϵ_i of the Schrödinger equation, the corresponding Green function is required for many classes of applications. Problems dealing with impurities, transport, disordered systems, photoemission, etc. are most naturally solved using the Green function. In principle, the Green function can be constructed from eigenfunctions, using the spectral representation,

$$G(\mathbf{x}, \mathbf{x}') = \sum_i \psi_i(\mathbf{x})(\epsilon - \epsilon_i)^{-1}\psi_i^*(\mathbf{x}').$$

This representation converges very slowly in general, and an alternative method is needed. Direct methods for computing the Green function are considered here, using boundary-matching methods of multiple scattering theory. An important consideration, in the context of extensions and applications of density functional theory (DFT), is that any external perturbing potential will in general be modified or screened by a self-consistent response. Hence any incremental potential considered here is assumed to be modified or screened in the sense of linear-response theory [79, 292].

7.5.1 Definitions

For energy κ^2 in Rydberg units, the Green function of the Helmholtz equation, with given boundary conditions, satisfies

$$(\nabla^2 + \kappa^2)G_0(\mathbf{x}, \mathbf{x}') = \delta(\mathbf{x}, \mathbf{x}'), \tag{7.22}$$

while the corresponding Schrödinger Green function satisfies

$$(\nabla^2 + \kappa^2 - v(\mathbf{x}))G(\mathbf{x}, \mathbf{x}') = \delta(\mathbf{x}, \mathbf{x}'). \tag{7.23}$$

These two Green functions are related by the Lippmann–Schwinger integral equation

$$G(\mathbf{x}, \mathbf{x}') = G_0(\mathbf{x}, \mathbf{x}') + \int_{\Re^3} G_0(\mathbf{x}, \mathbf{y})v(\mathbf{y})G(\mathbf{y}, \mathbf{x}')d^3\mathbf{y}. \tag{7.24}$$

The wave function of a perturbed system defined by $\Delta v(\mathbf{x})$ is

$$\tilde{\psi}(\mathbf{x}) = \psi(\mathbf{x}) + \int_{\Re^3} G(\mathbf{x}, \mathbf{x}')\Delta v(\mathbf{x}')\tilde{\psi}(\mathbf{x}')d^3\mathbf{x}'. \tag{7.25}$$

This equation can be applied to a wide class of problems including the electronic structure of vacancies, impurities, and of other localized perturbations of solids and atomic clusters. The methods considered here make it possible to construct $G(\mathbf{x}, \mathbf{x}')$ for any potential function $v(\mathbf{x})$ defined throughout the coordinate space $\Re^3$.

It is assumed that $\Re^3$ is subdivided into space-filling cells. A typical cell τ_μ has closed surface σ_μ. An external empty cell can be included to fill space for a finite system. Solutions of the homogeneous Helmholtz equation about the origin of cell μ define regular and irregular generalized solid-harmonic functions, denoted here by J_L^μ and N_L^μ, respectively. Each generalized solid-harmonic function is the product of a radial factor and a spherical-harmonic function, defined here by a combined index L that denotes the quantum numbers (ℓm). Normalization is assumed to be consistent with the Kronecker-delta relation,

$$(N_L|W_\sigma|J_{L'}) = \delta_{L,L'}. \tag{7.26}$$

Expressed in local coordinates for each atomic cell, the Helmholtz Green function can be expanded in solid-harmonic functions,

$$G_0(\mathbf{r}, \mathbf{r}') = -\sum_L J_L(\mathbf{r})N_L^*(\mathbf{r}'), \qquad r < r'. \tag{7.27}$$

To verify this expansion, consider $G_0(\mathbf{r}, \mathbf{r}')$ for a fixed point $\mathbf{r}'$. For $r < r'$, this function is a regular solution of the Helmholtz equation and must have an expansion of the form

$$G_0(\mathbf{r},\mathbf{r}') = \sum_L J_L(\mathbf{r})(N_L|W_0|G_0), \tag{7.28}$$

where the Wronskian integral is over the surface of any sphere centered at the cell origin with radius $r_0 < r'$. For $r > r'$, this function is a bounded solution of the Helmholtz equation and must have an expansion of the form

$$G_0(\mathbf{r},\mathbf{r}') = -\sum_L N_L(\mathbf{r})(J_L|W_1|G_0), \tag{7.29}$$

where the Wronskian integral is over the surface of any sphere centered at the cell origin with radius $r_1 > r'$. If Eq. (7.22) is multiplied by $N_L^*(\mathbf{r})$, and the corresponding null term with the Helmholtz operator applied to N_L^* is subtracted, on integrating over the volume defined by $r_0 < r' < r_1$, between radii r_0 and r_1, this integral reduces to Wronskian surface integrals. This implies

$$N_L^*(\mathbf{r}') = (N_L|W_1|G_0) - (N_L|W_0|G_0). \tag{7.30}$$

Since $r_1 > r'$, Eq. (7.29) implies that the first integral on the right-hand side vanishes. The second integral gives the coefficient in Eq. (7.28), which reduces to Eq. (7.27). Similarly, if Eq. (7.22) is multiplied by $J_L^*(\mathbf{r})$ and integrated over the volume between radii r_0 and r_1,

$$J_L^*(\mathbf{r}') = (J_L|W_1|G_0) - (J_L|W_0|G_0). \tag{7.31}$$

In this case, the second integral on the right-hand side vanishes and the first integral gives the coefficient in Eq. (7.29). This provides a formula valid for $r > r'$, complementary to Eq. (7.27). Comparison of these two formulas shows that the linear operator G_0 is Hermitian for real energies.

When coordinate $\mathbf{x} = \mathbf{X}_\mu + \mathbf{r}$ lies in cell τ_μ and $\mathbf{x}' = \mathbf{X}_\nu + \mathbf{r}'$ lies in a different cell τ_ν, a two-center expansion of the Green function is needed. Since $N_L^\mu(\mathbf{x}' - \mathbf{X}_\mu)$ is a regular solution of the Helmholtz equation in a sphere of radius $|\mathbf{X}_\mu - \mathbf{X}_\nu|$ about the origin of a displaced cell ν,

$$N_L^\mu(\mathbf{r}' + \mathbf{X}_\nu - \mathbf{X}_\mu) = N_L^\mu(\mathbf{r}')\delta_{\mu\nu} - \sum_{L'} J_{L'}^\nu(\mathbf{r}') g_{L',L}^{\nu\mu}, \tag{7.32}$$

for $r' < |\mathbf{X}_\mu - \mathbf{X}_\nu|$. The expansion coefficients $g_{L',L}^{\nu\mu}$ are structure constants at the given energy. This matrix is Hermitian for real energies. Using the above expansion, the Helmholtz Green function can be written in the form,

$$G_0(\mathbf{X}_\mu + \mathbf{r}, \mathbf{X}_\nu + \mathbf{r}') = \sum_{LL'} J_L^\mu(\mathbf{r}) g_{LL'}^{\mu\nu} J_{L'}^{\nu *}(\mathbf{r}') - \sum_L J_L^\mu(\mathbf{r}) N_L^{\nu *}(\mathbf{r}')\delta_{\mu\nu}, \tag{7.33}$$

where $r < r'$ within any single cell μ. The two-center expansion converges in general for muffin-tin geometry, when the two coordinate values lie within nonoverlapping spheres. It is assumed here that cell coordinate origins are chosen to satisfy this geometric condition for some neighborhood of each cell origin.

These definitions can be generalized to give an expansion of the Green function $G(\mathbf{x}, \mathbf{x}')$ in terms of local solutions of the Schrödinger equation in each cell τ. Local regular solutions ϕ_L and local irregular solutions ζ_L can be defined by matching to functions J_L and N_L, respectively, on an infinitesimal sphere about the cell origin, or, alternatively, by requiring similar conditions on a local cell boundary σ. This construction imposes the Wronskian condition

$$(\zeta_L|W_\sigma|\phi_{L'}) = \delta_{L,L'}. \tag{7.34}$$

These integrals vanish between any two ϕ or two ζ functions. Green's theorem implies that this Kronecker-delta formula is valid for Wronskian integrals over any closed surface that encloses the local cell center and excludes all other singular points of the potential function. If singularities of the potential function occur only at cell centers, these functions can be assumed to be have regular extensions out to the nearest neighbor cell center, but in general no further. For this reason, the primitive function ζ must be modified by the addition of a sum of regular functions ϕ in its cell of origin to define a modified irregular function $\tilde{\zeta}$ that is regular everywhere except at the defining cell center. With these definitions [53, 441, 444], by analogy to the Helmholtz Green function,

$$G(\mathbf{r}, \mathbf{r}') = -\sum_L \phi_L(\mathbf{r})\tilde{\zeta}^*_L(\mathbf{r}'), \qquad r < r', \tag{7.35}$$

$$\tilde{\zeta}^\mu_L(\mathbf{r}' + \mathbf{X}_\nu - \mathbf{X}_\mu) = \zeta^\mu_L(\mathbf{r}')\delta_{\mu\nu} - \sum_{L'} \phi^\nu_{L'}(\mathbf{r}')G^{\nu\mu}_{L',L}, \tag{7.36}$$

for $r' < |\mathbf{X}_\mu - \mathbf{X}_\nu|$, and

$$G(\mathbf{X}_\mu + \mathbf{r}, \mathbf{X}_\nu + \mathbf{r}') = \sum_L \sum_{L'} \phi^\mu_L(\mathbf{r})G^{\mu\nu}_{L,L'}\phi^{\nu*}_{L'}(\mathbf{r}') - \sum_L \phi^\mu_L(\mathbf{r})\zeta^{\nu*}_L(\mathbf{r}')\delta_{\mu\nu}, \tag{7.37}$$

valid for $r < r'$ within a single cell, or for coordinates in two nonoverlapping spheres. The coefficients $G^{\mu\nu}_{L,L'}$ define the structural matrix of the Schrödinger Green function. For $\nu = \mu$, $\tilde{\zeta}^\mu - \zeta^\mu$ is a sum of regular functions ϕ^μ with coefficients $-G^{\mu\mu}_{L,L'}$.

The definition of irregular functions ζ depends on boundary conditions imposed on the Green function. Any variation of the set of functions ζ defined by adding linear combinations of the regular functions ϕ with coefficients that constitute an Hermitian matrix simply moves the corresponding term between the two summations in Eq. (7.37), leaving the net sum invariant, and preserving the Kronecker-delta conditions of Eq. (7.34). The present derivations are covariant with respect to the group of such transformations. For particular applications, the representation can be chosen for the greatest convenience, subject only to the Kronecker-delta condition.

7.5.2 Properties of the Green function

The definition of the Schrödinger Green function can be extended to a complex-valued energy parameter z. Then Eq. (7.23) for the Schrödinger equation can be written as

$$(z - \mathcal{H}(\mathbf{x}))G(z; \mathbf{x}, \mathbf{x}') = \delta(\mathbf{x}, \mathbf{x}'), \tag{7.38}$$

which defines $G(z) = (z - \mathcal{H})^{-1}$ as the resolvent operator. $G(z)$ is an analytic operator function in the upper half of the complex z-plane. The singularities are poles or a continuum branch cut corresponding to energy eigenvalues. If $\mathcal{H}$ is Hermitian, the eigenvalues lie on the real energy axis. The corresponding orthonormal set of eigenfunctions ψ_i defines a spectral representation

$$G(z; \mathbf{x}, \mathbf{x}') = \sum_i \psi_i(\mathbf{x})(z - \epsilon_i)^{-1}\psi_i^*(\mathbf{x}'). \tag{7.39}$$

This formula indicates that the residue of $G(z)$ at an eigenvalue pole is just the density matrix, in the coordinate representation, summed over all eigenfunctions with this eigenvalue. This residue can be extracted conveniently by using Dirac's formula after displacing each eigenvalue into the lower complex plane,

$$\lim_{\delta\to 0}\frac{1}{z - \epsilon_i + i\delta} = \mathcal{P}\frac{1}{z - \epsilon_i} - i\pi\delta(z - \epsilon_i). \tag{7.40}$$

Specializing to the electronic density for electrons of one spin in the local coordinates of a particular cell, the local density per unit energy is

$$n(\epsilon; \mathbf{r}) = -\frac{1}{\pi}\Im G(\epsilon; \mathbf{r}, \mathbf{r}). \tag{7.41}$$

Summing over all eigenenergies below a chemical potential or Fermi level μ, the local density function is defined by a contour integral passing above all poles on the real axis,

$$\rho(\mathbf{r}) = -\frac{1}{\pi}\int_{-\infty}^{\mu}\Im G(z; \mathbf{r}, \mathbf{r})\,dz. \tag{7.42}$$

In methodology based on computation of the Green function, this formula replaces the usual sum over eigenfunctions. When integrated over coordinate space,

Eq. (7.41) gives the density of states per unit energy,

$$n(\epsilon) = -\frac{1}{\pi} \int_{\mathcal{R}^3} \Im G(\epsilon; \mathbf{r}, \mathbf{r}) \, d\tau. \quad (7.43)$$

These formulas must be summed over a spin index to give the corresponding total densities. In multiple scattering theory, the Green function as given by Eq. (7.37) is subdivided into terms valid in separate atomic cells. The elements with $\mathbf{r}' = \mathbf{r}$ are single sums over the cells. Hence the density of states of given spin in a particular cell τ_μ is

$$n^\mu(\epsilon) = -\frac{1}{\pi} \int_{\tau_\mu} \Im G^\mu(\epsilon; \mathbf{r}, \mathbf{r}), \quad (7.44)$$

where G^μ denotes the terms with $\nu = \mu$ in Eq. (7.37).

7.5.3 Construction of the Green function

Equation (7.24) suggests that the structural matrix $G^{\mu\nu}_{L,L'}$ of the Schrödinger Green function at specified energy should have a simple relationship to that of the Helmholtz Green function, which is the matrix of structure constants $g^{\mu\nu}_{L,L'}$. The basis set of regular local solutions of the Schrödinger equation can be constructed so that matrix G depends only on g and the t-matrix. In deriving the KKR/MST equations, primitive basis functions were defined by outward integration, starting from specified regular solid-harmonic functions J_L at the origin in each atomic cell. A canonical basis set is defined by transforming the primitive basis by matrix C^{-1}, where $C_{L,L'}$ is defined as in MST. Hence, in the canonical basis, C is a unit matrix. The matching equations on the local cell boundary depend only on the t-matrix, because each basis function matches to a boundary function $J - Nt$. These basis functions, still indexed by L, are in one-to-one correspondence with the regular solid-harmonic functions J_L. Corresponding to the canonical regular basis function ϕ_L, which matches onto $J - Nt$ on cell surface σ, a paired canonical irregular function ζ_L can be computed by integrating the Schrödinger equation inwards from σ, starting from a single irregular function N_L on this surface. The full set of canonical basis functions satisfies Eq. (7.34).

The structural matrix G can be derived in this canonical basis from the matrix g of structure constants, using the surface-matching theorem. For a modified irregular function $\tilde{\zeta}^\lambda$ and the corresponding global matching function $\xi = \Sigma_\mu N^\mu \beta^\mu$, the matching conditions on the boundary of any cell τ_μ are

$$(J^\mu_L | W^\mu_\sigma | \tilde{\zeta}^\lambda - \xi) = 0, \qquad (N^\mu_L | W^\mu_\sigma | \tilde{\zeta}^\lambda - \xi) = 0. \quad (7.45)$$

Expanding the difference function $\tilde{\xi}^\lambda - \zeta^\lambda$ in the form $-\phi^\mu G^{\mu\lambda}$ in cell τ_μ, its representation in the canonical basis on surface σ_μ is $-(J^\mu - N^\mu t^\mu)G^{\mu\lambda}$. Similarly, ζ^λ is $N^\mu \delta^{\mu\lambda}$ on σ_μ and ξ is $N^\mu \beta^\mu - \Sigma_\nu J^\mu g^{\mu\nu} \beta^\nu$. Then from Eqs. (7.45) the coefficients in functions ξ and $\tilde{\xi}^\lambda - \zeta^\lambda$, respectively, are given by

$$\beta^\mu = \delta^{\mu\lambda} + t^\mu G^{\mu\lambda}, \qquad G^{\mu\lambda} = \sum_\nu g^{\mu\nu} \beta^\nu, \tag{7.46}$$

omitting indices L. Combining these equations, for each cell τ_μ,

$$G^{\mu\lambda} = g^{\mu\lambda} + \sum_\nu g^{\mu\nu} t^\nu G^{\nu\lambda}, \tag{7.47}$$

again omitting L-indices and summations. This is a Dyson equation in the form of a matrix representation $G = g + gtG$ of the Lippmann–Schwinger equation, Eq. (7.24). This equation can also be derived by using the idea of atomic-cell orbitals (ACO), generalized to irregular functions. An irregular ACO is defined in a reference cell τ_λ as the canonical function ζ^λ. In all other cells τ_μ it is the local expansion of N^λ, expressed in terms of the structure constants $g^{\mu\lambda}$. Similarly, a regular ACO takes the form $\phi - J$ in its indexed cell and Jgt in all other cells. The ACO form of $\tilde{\xi}^\lambda$ in cell τ_μ is

$$\tilde{\xi}^\lambda = \zeta^\mu \delta^{\mu\lambda} - \phi^\mu G^{\mu\lambda} + J^\mu \left[\sum_\nu (\delta^{\mu\nu} - g^{\mu\nu} t^\nu) G^{\nu\lambda} - g^{\mu\lambda} \right], \tag{7.48}$$

omitting indices L. Following tail-cancellation logic [7], the coefficients of functions J must vanish in all cells. This implies Eq. (7.47).

The structural matrix $G^{\mu\nu}_{L,L'}$ can be evaluated by direct matching across adjacent cell interfaces, without using the matrix of structure constants. This construction was originally derived (unpublished notes) from the Green function cellular method (GFCM) [53, 441, 444]. The GFCM itself can be derived as a variant of the variational cellular method (VCM) [112, 110, 276], by restricting the form of allowed variations on cell surfaces. The resulting NVCM or GFCM equations are given above as Eqs. (7.6).

The structure constants or structural matrix for the Helmholtz Green function, defined by Eq. (7.22), can be determined by applying Eqs. (7.6) to the global solution of the Helmholtz equation given by Eq. (7.32), with a specified singularity in cell τ_λ. For this application, the potential function vanishes everywhere except for an implied singular potential in cell τ_λ, designed to specify the irregular solid harmonic function N^λ_L as an exact solution in this cell. Then the local basis in each cell $\nu \neq \lambda$ is the set of regular solid harmonic functions J^ν_L. Using these local expansions, Eqs. (7.6) become a set of inhomogeneous linear equations for column λ of the matrix of structure constants. Suppressing L-indices, these equations take the form,

$$\sum_{\nu \neq \mu} (N^{\mu}|W_{\mu\nu}|(N^{\mu}\delta^{\mu\lambda} - J^{\mu}g^{\mu\lambda}) - (N^{\nu}\delta^{\nu\lambda} - J^{\nu}g^{\nu\lambda})) = 0, \quad (7.49)$$

for all cells τ_{μ}. Simplifying by use of Eq. (7.26),

$$\sum_{\nu \neq \mu} (N^{\mu}|W_{\mu\nu}|J^{\nu})g^{\nu\lambda} - g^{\mu\lambda} = \sum_{\nu \neq \mu} (N^{\mu}|W_{\mu\nu}|N^{\nu})\delta^{\nu\lambda}. \quad (7.50)$$

While these equations are consistent with the GFCM, their relationship to the VCM variational equations suggests that the alternative formalism defined by the JVCM might be more appropriate. In the global solution sought in the present case, the irregular term is fixed, and all variational increments are linear combinations of regular functions J_L. Hence variation of coefficients gives the JVCM Eqs. (7.7). A similar situation occurs for the Poisson equation, since a particular solution of the inhomogeneous equation is fixed, and all incremental functions are regular solid harmonics. Recent calculations comparing these methods for the Poisson equation [443] show improved convergence in the JVCM formalism. It was shown that extension of the Schlosser–Marcus variational principle to the Poisson equation implies the JVCM equations [443]. In the present case, Eqs. (7.7) take the form

$$\sum_{\nu \neq \mu} (J^{\mu}|W_{\mu\nu}|(N^{\mu}\delta^{\mu\lambda} - J^{\mu}g^{\mu\lambda}) - (N^{\nu}\delta^{\nu\lambda} - J^{\nu}g^{\nu\lambda})) = 0, \quad (7.51)$$

for all cells τ_{μ}. Simplifying by use of Eq. (7.26),

$$\sum_{\nu \neq \mu} (J^{\mu}|W_{\mu\nu}|J^{\nu})g^{\nu\lambda} = \sum_{\nu \neq \mu} (J^{\mu}|W_{\mu\nu}|N^{\nu})\delta^{\nu\lambda} + \delta^{\mu\lambda}. \quad (7.52)$$

Equation (7.52) is generalized to the Schrödinger Green function by using local regular functions ϕ on the left and by using Eq. (7.36) to define an exact solution of the Schrödinger equation in each cell, with a specified singularity in cell τ_{λ}. The resulting linear equations take the form, suppressing L-indices,

$$\sum_{\nu \neq \mu} (\phi^{\mu}|W_{\mu\nu}|\phi^{\nu})G^{\nu\lambda} = \sum_{\nu \neq \mu} (\phi^{\mu}|W_{\mu\nu}|\zeta^{\nu})\delta^{\nu\lambda} + \delta^{\mu\lambda}, \quad (7.53)$$

for all values of μ and λ. These equations determine the structural matrix G for any set of space-filling cells.

Equations (7.53) simplify if the system has point-group or translational symmetry. Then local basis functions of equivalent cells are related to those of a smaller number of generating cells by phase factors and elementary rotations. Considering only translational symmetry, it is convenient to index cells in a reference translational cell by indices μ, etc. and to index translated equivalent cells by the corresponding indices $\tilde{\mu}$, etc., such that the displacement of cell $\tau_{\tilde{\mu}}$ relative to cell τ_{μ}

is a lattice translational vector $\mathbf{d}_{\tilde{\mu}\mu}$. Using this notation, Eqs. (7.53) imply separate equations for each **k**-vector in the reduced Brillouin zone, in the form

$$\sum_{\tilde{\nu}\neq\mu}(\phi^{\mu}|W_{\mu\tilde{\nu}}|\phi^{\tilde{\nu}})\exp(i\mathbf{k}\cdot\mathbf{d}_{\tilde{\nu}\nu})G^{\nu\lambda}(\mathbf{k}) =$$
$$\sum_{\tilde{\nu}\neq\mu}(\phi^{\mu}|W_{\mu\tilde{\nu}}|\zeta^{\tilde{\nu}})\exp(i\mathbf{k}\cdot\mathbf{d}_{\tilde{\nu}\nu})\delta^{\nu\lambda} + \delta^{\mu\lambda}, \qquad (7.54)$$

for all values of μ and λ that index cells in the reference translational cell. The corresponding equations derived from Eq. (7.50) have been verified by computing structure constants for an fcc lattice along the $\Gamma - K$ line in the reduced Brillouin zone, checked against a published structure-constant program STR [384]. Equations (7.53) have not been tested, but are expected to show improved convergence because they are variationally correct for this problem.

8

Variational methods for continuum states

The principal references for this chapter are:

[3] Adhikari, S.K. (1998). *Variational Principles and the Numerical Solution of Scattering Problems* (Wiley, New York).

[47] Burke, P.G. and Robb, W.D. (1975). The R-matrix theory of atomic processes, *Adv. At. Mol. Phys.* **11**, 143–214.

[178] Huo, W.M. and Gianturco, F.A. (1995). *Computational Methods for Electron-Molecule Collisions* (Plenum, New York).

[202] Kohn, W. (1948). Variational methods in nuclear collision problems, *Phys. Rev.* **74**. 1763–1772.

[214] Lane, A.M. and Thomas, R.G. (1958). R-matrix theory of nuclear reactions, *Rev. Mod. Phys.* **30**, 257–353.

[215] Lane, N.F. (1980). The theory of electron–molecule collisions, *Rev. Mod. Phys.* **52**, 29–119.

[246] Moiseiwitsch, B.L. (1966). *Variational Principles* (Interscience, New York).

[257] Mott, N.F. and Massey, H.S.W. (1965). *The Theory of Atomic Collisions* (Oxford University Press, New York).

[270] Nesbet, R.K. (1980). *Variational Methods in Electron-Atom Scattering Theory* (Plenum, New York).

[371] Schwinger, J. (1947). A variational principle for scattering problems, *Phys. Rev.* **72**, 742.

[428] Wigner, E.P. and Eisenbud, L. (1947). Higher angular momenta and long range interaction in resonance reactions, *Phys. Rev.* **72**, 29–41.

8.1 Scattering by an N-electron target system

Scattering by an N-electron atom or molecule with fixed nucleus is described by an $(N+1)$-electron Schrödinger wave function of the form

$$\Psi_s = \sum_p \mathcal{A}\Theta_p \psi_{ps} + \sum_\mu \Phi_\mu c_{\mu s}. \tag{8.1}$$

The index s here denotes a particular degenerate solution, specified by boundary conditions, at given total energy E. The N-electron function Θ_p is a target state with energy E_p. It is assumed that all target functions Θ_p are defined by orthonormal eigenvectors of a Hamiltonian matrix in a common set of N-electron basis functions. The eigenvalues are the energies E_p. ψ_{ps} is a one-electron *channel orbital* wave function, antisymmetrized into Θ_p by the operator $\mathcal{A}$. The functions Φ_μ are an orthonormal set of $(N+1)$-electron Slater determinants. The channel orbital ψ_{ps} is characterized by orbital and spin angular momentum quantum numbers and by asymptotic energy $E - E_p$. This energy is positive for *open* scattering channels, defining a wave vector or electron momentum k_p such that $E - E_p = \frac{1}{2}k_p^2$. Hartree atomic units are used here. If $E - E_p < 0$ the channel is *closed*. Then k_p is replaced by $i\kappa_p$, for $\kappa_p > 0$. In practice, the basis functions Φ_μ are replaced by symmetry-adapted linear combinations of simple Slater determinants, defined for total spin and atomic orbital angular momentum or for molecular point-group symmetry. The expansion indicated in Eq. (8.1) remains completely general. For heavy atoms, the notation can be adapted to jj-coupling.

For an atom, or outside a sphere that completely encloses a target molecule, channel orbital functions are of the form

$$\psi_{ps} = r^{-1} f_{ps}(r) Y_{\ell m_\ell}(\theta, \phi) u_{m_s},$$

where, for $r \to \infty$, $f_{ps}(r) \sim k_p^{-\frac{1}{2}} \sin(k_p r - \frac{1}{2}\ell_p \pi + \eta_p)$, for single-channel scattering by a neutral target. This functional form must be modified for long-range Coulomb and dipole potentials. The normalization implies unit radial flux density for a free electron. For interacting scattering channels, the multichannel wave functions are

$$f_{ps}(r) \sim k_p^{-\frac{1}{2}} \left[\sin\left(k_p r - \frac{1}{2}\ell_p \pi\right) \alpha_{0ps} + \cos\left(k_p r - \frac{1}{2}\ell_p \pi\right) \alpha_{1ps} \right]. \quad (8.2)$$

The coefficient matrices α_0 and α_1 determine scattering matrices and cross sections. Closed-channel orbital functions must vanish for $r \to \infty$, in a way determined by the long-range potentials in the Schrödinger equation. The term in $\exp(-\kappa_p r)$ implied by analytic continuation of an open-channel orbital through the continuum threshold is generally dominated by terms in reciprocal powers of r due to such long-range potentials.

It is assumed that ψ_{ps} is orthogonal to all orbital functions used to construct Θ_p and Φ_μ. This ensures that Eq. (8.1) describes projection of Ψ_s onto orthogonal target-state components $\Psi_p = \mathcal{A}\Theta_p\psi_{ps}$ and a residual orthogonal component $\Psi_Q = \sum_\mu \Phi_\mu c_{\mu s}$ that is quadratically integrable. The coefficients $c_{\mu s}$ are determined variationally. For any calculation using a finite orbital basis, the open-channel terms $\mathcal{A}\Theta_p\psi_{ps}$ remain distinct from the Hilbert-space component Ψ_Q. The oscillatory function ψ_{ps} with nonvanishing asymptotic amplitude cannot be represented as a finite superposition of quadratically integrable functions. In contrast, closed-channel terms can be included in either sum. In the *close-coupling* formalism, part of Ψ_Q can be replaced by $\mathcal{A}\Theta_\gamma\psi_{\gamma s}$, where Θ_γ represents a target *pseudostate* with energy $E_\gamma > E$, corresponding to a closed channel. Such a function lies in the electronic continuum. Polarization response can be represented in terms of such *polarization pseudostates* $\Theta_{\gamma(p)}$, computed as the first-order perturbation of Θ_p by a polarizing field. The corresponding closed-channel orbital function $\psi_{\gamma s}$ may be computed from the close-coupling equations derived here.

Using the projection-operator formalism of Feshbach [115, 116], an implicit variational solution for the coefficients $c_{\mu s}$ in Ψ_Q can be incorporated into an equivalent partitioned equation for the channel orbital functions. This is a multichannel variant of the logic used to derive the correlation potential operator $\hat{v}_c$ in orbital-functional theory. Define a projection operator Q such that

$$\Psi_Q = Q\Psi_s = \sum_\mu \Phi_\mu(\Phi_\mu|\Psi_s),$$

and the orthogonal complement operator P, such that

$$\Psi_P = P\Psi_s \simeq \sum_p \mathcal{A}\Theta_p\psi_{ps}.$$

Defining a Schrödinger functional $\Xi = (\Psi|H - E|\Psi)$, the Euler–Lagrange equations for fixed energy E are

$$\frac{\delta\Xi}{\delta\Psi_P^*} = 0; \qquad \frac{\delta\Xi}{\delta\Psi_Q^*} = 0.$$

The resulting coupled $(N+1)$-electron equations are

$$P(H-E)P\Psi_P + PHQ\Psi_Q = 0$$
$$QHP\Psi_P + Q(H-E)Q\Psi_Q = 0.$$

The reduced operator $Q(H-E)Q$ in the Q-space can be inverted to give a formal solution of the second equation,

$$\Psi_Q = -\{Q(H-E)Q\}^{-1}QHP\Psi_P.$$

The operator $\{Q(H-E)Q\}^{-1}$ is an $(N+1)$-electron linear operator whose kernel is

$$\sum_{\mu,\nu} \Phi_\mu(H-E)^{-1}_{\mu\nu}\Phi^*_\nu.$$

When substituted into the first equation, this gives an effective multichannel Schrödinger equation of the form

$$\{\tilde{H} - E\}\Psi_P = \{P(H-E)P - PHQ\{Q(H-E)Q\}^{-1}QHP\}\Psi_P = 0. \qquad (8.3)$$

Equation (8.3) provides a common basis for the major computational methods used in low-energy electron scattering theory. The terms containing $\{Q(H - E)Q\}^{-1}$ in this equation describe polarization response and other correlation effects in terms of a matrix optical potential. Singularities due to this energy denominator cancel out exactly [264, 265, 270].

Effective one-electron equations for the channel orbital functions can be obtained either by evaluating orbital functional derivatives of the variational functional Ξ or more directly by projecting Eq. (8.3) onto the individual target states Θ_p. With appropriate normalizing factors, $(\Theta_p|\Psi_s) = \psi_{ps}$. Equations for the radial channel functions $f_{ps}(r)$ are obtained by projecting onto spherical harmonics and elementary spin functions. The matrix operator acting on channel orbitals is

$$\hat{m}^{pq} = (\Theta_p|\tilde{H} - E|\mathcal{A}\Theta_q). \tag{8.4}$$

The projection of this operator onto spherical harmonics and spin functions defines a radial operator $m^{pq} = \hat{g}^{pq} - \epsilon_p\delta_{pq}$, where $\epsilon_p = E - E_p$.

8.1.1 Cross sections

If there are n_o open channels at energy E, there are n_o linearly independent degenerate solutions of the Schrödinger equation. Each solution Ψ_s is characterized by a vector of coefficients α_{ips}, for $i = 0, 1$, defined by the asymptotic form of the multichannel wave function in Eq. (8.1). The rectangular column matrix α consists of the two $n_o \times n_o$ coefficient matrices α_0, α_1. Any nonsingular linear combination of the column vectors of α produces a physically equivalent set of solutions. When multiplied on the right by the inverse of the original matrix α_0, the transformed α-matrix takes the canonical form

$$\alpha_0 = I, \qquad \alpha_1 = K,$$

which defines the *reactance* matrix $K = \alpha_1(\alpha_0)^{-1}$. For exact solutions of the projected Schrödinger equation, the open-channel K-matrix is real and symmetric [257, 302]. It can be diagonalized by an orthogonal transformation to define *eigenchannels*, defined by the eigenvectors. The eigenvalues are $\tan\eta_\sigma$, for $\sigma = 1, \ldots, n_o$, which defines the *eigenphases* η_σ up to a multiple of π radians. The *scattering* matrix S is defined by [257]

$$S = (I + iK)(I - iK)^{-1},$$

and the *transition* matrix T is defined by

$$T = \frac{1}{2i}(S - I) = K(I - iK)^{-1}.$$

In terms of eigenchannels and eigenphases,

$$K_{ps} = \sum_{\sigma=1}^{n_o} x_{p\sigma} x_{s\sigma} \tan \eta_\sigma$$

$$S_{ps} = \sum_{\sigma=1}^{n_o} x_{p\sigma} x_{s\sigma} \exp(2i\eta_\sigma)$$

$$T_{ps} = \sum_{\sigma=1}^{n_o} x_{p\sigma} x_{s\sigma} \exp(i\eta_\sigma) \sin(\eta_\sigma).$$

The S-matrix is unitary and symmetric, while the T-matrix is symmetric. This particular definition of the T-matrix reduces for scattering by a central potential to the phase-shift factor in the scattering amplitude,

$$f(\theta) = \frac{1}{k} \sum_{\ell=0}^{\infty} (2\ell + 1) \exp(i\eta_\ell) \sin \eta_\ell P_\ell(\cos\theta),$$

such that the differential cross section is

$$\frac{d\sigma}{d\Omega} = |f(\theta)|^2.$$

If k is in atomic units a_0^{-1}, the differential cross section is in units a_0^2 per steradian. Differential and total cross sections for multichannel scattering by an atomic target can be derived from general formulas [27, 184]. The partial cross section for scattering from channel q to channel p is

$$\sigma_{qp} = \frac{4\pi}{k_q^2} |T_{pq}|^2 = \frac{4\pi}{k_q^2} \left| \sum_\sigma x_{p\sigma} x_{q\sigma} \exp(i\eta_\sigma) \sin(\eta_\sigma) \right|^2.$$

The total cross section for unpolarized scattering is obtained by summing σ_{qp} over degenerate final states and by averaging over initial states.

8.1.2 Close-coupling expansion

If the set of target states Θ_p in Eq. (8.1) could be extended to completeness, solution of the coupled equations for the channel orbitals ψ_{ps} would provide a quantitative solution of the scattering problem for an N-electron target. This is the strategy of the close-coupling method [374, 325, 48, 376]. The coupled integrodifferential equations for the channel orbitals are the *close-coupling* equations. Because the expansion in target states must be truncated for any practical calculation, the second term in Eq. (8.1) is required in order to include polarization and correlation effects as well as to assure consistent orbital orthogonality conditions. No generally satisfactory way has been found to include the continuum of unbound target states required for completeness of this expansion.

Since only a relatively small number of coupled equations can be solved in practice, the target states Θ_p must be selected carefully. Virtual target excitation into the ionization continuum must be approximated by inclusion of closed-channel pseudostates that cannot be target eigenstates but have the character of wave packets in the continuum. Target polarization response is treated by including polarization pseudostates $\Theta_{\gamma(p)}$.

If only a single target state is included, itself represented as a single-determinant model state Φ_p, the formalism reduces to the *static-exchange* (SE) model. Because of the complexity of more detailed scattering models, much work in fixed-nuclei electron scattering by molecules has been carried out at this level of approximation [215]. This model is generally unsatisfactory for low-energy scattering, and must be augmented by a valid approximation to the dipole polarization response of the target system [270, 135]. Models that augment static exchange with polarization response are denoted by the acronym SEP. It will be shown below that a model with this computational structure can be derived from formally exact orbital functional theory (OFT), if the correlation potential operator $\hat{v}_c$ is approximated in terms of polarization pseudostates [290].

8.2 Kohn variational theory

Variational theory relevant to the representation of an $(N+1)$-electron continuum wave function by Eq. (8.1) will be considered here. Direct computational implementation of this formalism defines the *matrix variational method*. Since the widely applied close-coupling method can be described by Eq. (8.1), the general argument here also applies to that formalism, with some modifications of detail. In the *algebraic close-coupling* method, the variational formalism, using an orbital basis expansion, replaces direct numerical solution of integrodifferential close-coupling equations.

In bound-state calculations, the Rayleigh–Ritz or Schrödinger variational principle provides both an upper bound to an exact energy and a stationary property that determines free parameters in the wave function. In scattering theory, the energy is specified in advance. Variational principles are used to determine the wave function but do not generally provide variational bounds. A variational functional is made stationary by choice of variational parameters, but the sign of the residual error is not determined. Because there is no well-defined bounded quantity, there is no simple absolute standard of comparison between different variational trial functions. The present discussion will develop a stationary estimate of the multichannel K-matrix. Because this matrix is real and symmetric for open channels, it provides the most convenient of several alternative representations of the asymptotic part of a scattering wave function. Recent developments that exploit aspects of the complex-valued *S*- or *T*-matrices will be discussed below.

Variational principles for scattering matrices and wave functions were originally introduced by Hulthén [175, 176] and by Kohn [202]. Quantitative applications of variational methods to electron–atom scattering theory began with calculations of e^-–H elastic scattering by Schwartz [369, 370], using the Kohn method for the *K*-matrix. Further progress required understanding and resolution of the problem of spurious singularities in the Kohn formalism. These developments have been discussed in detail elsewhere [270]. More recently, this problem has largely been bypassed in the *complex Kohn* formalism, which has been very effectively used for calculations of electron–molecule scattering [341].

8.2.1 The matrix variational method

The asymptotic form of radial open-channel orbitals $f_{ps}(r)$ is given by Eq. (8.2). Functions of this form can be represented as linear combinations of two independent continuum basis functions for each open channel. These basis functions must be regular at the coordinate origin, but have the asymptotic forms

$$\begin{aligned} F_{0p}(r) &\sim k_p^{-\frac{1}{2}} \sin\left(k_p r - \frac{1}{2}\ell_p \pi\right), \\ F_{1p}(r) &\sim k_p^{-\frac{1}{2}} \cos\left(k_p r - \frac{1}{2}\ell_p \pi\right). \end{aligned} \tag{8.5}$$

A variational approximation with the required asymptotic form

$$f_{ps}(r) = F_{0p}\alpha_{0ps} + F_{1p}\alpha_{1ps} \tag{8.6}$$

determines the coefficients α_{ips} defined by Eq. (8.2). In the *matrix variational method* [270], the quadratically integrable orbital basis is augmented for each open channel by adding two functions ϕ_{0p}, ϕ_{1p} whose radial factors are $r^{-1}F_{0p}(r)$ and $r^{-1}F_{1p}(r)$, respectively. These functions are orthogonalized to the quadratically integrable orbital basis $\{\phi_i; \phi_a\}$.

For an exact scattering wave function, the coefficients $c_{\mu s}$ in Eq. (8.1) would be determined as linear functions of the matrix elements α_{ips}, since the algebraic equations are linear. By factoring these coefficients

$$c_{\mu s} = \sum_{ip} c_\mu^{ip} \alpha_{ips},$$

the variational approximation to the $(N+1)$-electron wave function takes the form

$$\Psi_s = \sum_{ip} \left(\mathcal{A}\Theta_p \phi_{ip} + \sum_{\mu} \Phi_\mu c_\mu^{ip} \right) \alpha_{ips}.$$

In the matrix variational method, the coefficients c_μ^{ip} are determined separately for each set of indices i, p from the matrix equations, for all μ,

$$\left(\Phi_\mu | H - E | \mathcal{A}\Theta_q \phi_{jq} + \sum_\nu \Phi_\nu c_\nu^{jq} \right) = 0, \tag{8.7}$$

which imply the variational condition $(\Phi_\mu|H - E|\Psi_s) = 0$. For arbitrary values of the coefficients α_{ips} these equations follow from the variational condition

$$\frac{\partial \Xi_{st}}{\partial c_\mu^{ip*}} = 0,$$

for all μ, i, p. The variational functional $\Xi_{st} = (\Psi_s|H - E|\Psi_t)$ here is an $n_o \times n_o$ matrix.

When Eq. (8.7) is satisfied, the variational functional becomes an explicit quadratic function of the coefficients α_{ips}, which can be assumed to be real numbers. Thus

$$\Xi_{st} = \sum_{ip} \sum_{jq} \alpha_{ips} m_{ij}^{pq} \alpha_{jqt},$$

where, as a consequence of Eq. (8.7),

$$m_{ij}^{pq} = M_{ij}^{pq} - \sum_{\mu,\nu} M_{ip,\mu} (M^{-1})_{\mu\nu} M_{\nu,jq}.$$

This equation combines several submatrices of $H - E$: the Hermitian *bound–bound* matrix $M_{\mu\nu}$, the rectangular *bound–free* matrix $M_{\mu,ip}$, and the nonhermitian *free–free* matrix M_{ij}^{pq}, where

$$\begin{aligned} M_{\mu\nu} &= (\Phi_\mu|H - E|\Phi_\nu), \\ M_{\mu,ip} &= (\Phi_\mu|H - E|\mathcal{A}\Theta_p\phi_{ip}), \\ M_{ij}^{pq} &= (\mathcal{A}\Theta_p\phi_{ip}|H - E|\mathcal{A}\Theta_q\phi_{jq}). \end{aligned}$$

Phase factors can be chosen so that both M_{ij}^{pq} and m_{ij}^{pq} are real but unsymmetrical.

If Ψ_t were an exact solution of the Schrödinger equation, and total energy E is above the thresholds for n_o open channels, there would be n_o linearly independent solutions of the equations

$$\sum_{jq} m_{ij}^{pq} \alpha_{jqt} = 0, \tag{8.8}$$

for all indices i, p. Each such solution defines a $2n_o$ element column vector of the rectangular matrix of coefficients α. In a matrix notation, suppressing channel indices, the variational functional is

$$\Xi = \alpha^\dagger m\alpha = \alpha_0^\dagger(m_{00}\alpha_0 + m_{01}\alpha_1) + \alpha_1^\dagger(m_{10}\alpha_0 + m_{11}\alpha_1).$$

In this notation, Eq. (8.8) is

$$m\alpha = \begin{pmatrix} m_{00} & m_{01} \\ m_{10} & m_{11} \end{pmatrix} \begin{pmatrix} \alpha_0 \\ \alpha_1 \end{pmatrix} = 0.$$

The reactance matrix K is $\alpha_1\alpha_0^{-1}$. Exact solutions require the matrix m_{ij}^{pq} to be of rank n_o, implying n_o linearly independent null-vectors as solutions of the homogeneous equations $m\alpha = 0$. Because this algebraic condition is not satisfied in general by approximate wave functions, a variational method is needed in order to specify in some sense an optimal approximate solution matrix α.

If α were an exact solution of the matrix equation $m\alpha = 0$, the functional $\Xi = \alpha^\dagger m\alpha$ would vanish. For a variational approximation, consider the variation of Ξ induced by an infinitesimal variation $\delta\alpha$. This is

$$\delta\Xi = \delta\alpha^\dagger m\alpha + (m\alpha)^\dagger\delta\alpha + \alpha^\dagger(m - m^\dagger)\delta\alpha.$$

The last term here does not vanish even if $m\alpha = 0$. The nonhermitian part of m_{ij}^{pq} comes from the free–free matrix of $H - E$, which is characteristic of scattering theory. When channel orbital functions are normalized to unit flux, with the asymptotic forms specified above, this nonhermitian term is given by

$$m_{ij}^{pq} - m_{ji}^{qp} = M_{ij}^{pq} - M_{ji}^{qp} = \frac{1}{2}\delta_{pq}(\delta_{i0}\delta_{j1} - \delta_{i1}\delta_{j0}).$$

This formula expresses the surface integral obtained by integrating the kinetic energy integral by parts for open-channel orbital functions of the specified asymptotic form. In matrix notation,

$$m_{01} - m_{10}^\dagger = \frac{1}{2}I,$$

where I is the $n_o \times n_o$ unit matrix. The matrices m_{00} and m_{11} are Hermitian (real and symmetric). When substituted into the expression for $\delta\Xi$ this gives

$$\delta\Xi = \delta\alpha^\dagger m\alpha + (m\alpha)^\dagger\delta\alpha + \frac{1}{2}(\alpha_0^\dagger\delta\alpha_1 - \alpha_1^\dagger\delta\alpha_0). \tag{8.9}$$

8.2.2 The Hulthén–Kohn variational principle

Although $\delta\Xi$ does not vanish even if the variational equation $m\alpha = 0$ is satisfied, the nonvanishing terms can be combined with variations of the various scattering matrices to give stationary expressions for these matrices. This logic produces multichannel versions of the variational principle originally derived by Hulthén [175, 176], Kohn [202], and Kato [194, 195]. As applied to the K-matrix, Kohn's variational principle can be derived by considering variations about an exact solution that maintain the canonical form $\alpha_0 = I, \alpha_1 = K$. Then $\delta\alpha_0 = 0$ and $\delta\alpha_1 = \delta K$, which implies $\delta\Xi = \frac{1}{2}\delta K$. This implies that the Kohn functional

$$[K] = K - 2\Xi$$

is stationary for all such variations.

In the matrix variational method, the equations $m\alpha = 0$ do not in general have a solution. For variations about an estimated matrix K_t, and restricted to the canonical form,

$$\delta\Xi = \delta K^\dagger(m_{10} + m_{11}K_t) + (m_{10} + m_{11}K_t)^\dagger \delta K + \frac{1}{2}\delta K.$$

If $K_t = -m_{11}^{-1}m_{10}$, such that $m_{10} + m_{11}K_t = 0$, this implies that $[K] = K_t - 2\Xi(K_t)$ is stationary. This defines

$$\begin{aligned}[K] &= K_t - 2(m_{00} + m_{01}K_t + K_t^\dagger m_{10} + K_t^\dagger m_{11}K_t) \\ &= -2\big(m_{00} - m_{10}^\dagger m_{11}^{-1} m_{10}\big).\end{aligned} \tag{8.10}$$

This is a real symmetric matrix by construction, and remains valid for an exact variational solution, when $\Xi = 0$. Hence this derivation proves that an exact open-channel K-matrix is real and symmetric.

The difficulty with this formula is that there is no guarantee that the auxiliary matrix m_{11} is not singular [264, 265]. At an energy for which an eigenvalue of m_{11} vanishes, K_t is not defined unless all columns of the matrix m_{10} are orthogonal to the null eigenvector. Specific examples [265] show that m_{10} does not have this property in general for otherwise valid variational trial wave functions. The number of null values of m_{11} increases with the number of basis functions, leading to energy and basis-dependent anomalies in the Kohn formalism [369, 370]. In an alternative version of the theory [176, 350], variations are restricted to $\alpha_0 = K^{-1}$ and $\alpha_1 = I$. This implies that anomalies occur at null points of m_{00}, which in general do not coincide with those of m_{11}. This fact can be exploited to generate various anomaly-free variants of the Kohn theory [265]. Especially in the matrix variational formalism, this method has been applied to many calculations of electron–atom scattering of high accuracy [270]. This method was used to compute e^-–He scattering cross sections that established a calibration standard for subsequent experimental work [268, 269].

8.2.3 The complex Kohn method

The derivation given above of the stationary Kohn functional $[K]$ depends on logic that is not changed if the functions F_0 and F_1 of Eq. (8.5) are replaced in each channel by any functions for which the Wronskian condition $m_{01} - m^{\dagger}_{10} = \frac{1}{2}I$ is satisfied [245, 191]. The *complex Kohn* method [244, 237, 440] exploits this fact by defining continuum basis functions consistent with the canonical form $\alpha_0 = I, \alpha_1 = T$, where T is the complex-symmetric multichannel transition matrix. These continuum basis functions have the asymptotic forms

$$u_{0p}(r) \sim k_p^{-\frac{1}{2}} \sin\left(k_p r - \frac{1}{2}\ell_p \pi\right),$$

$$u_{1p}(r) \sim k_p^{-\frac{1}{2}} \exp i\left(k_p r - \frac{1}{2}\ell_p \pi\right).$$

This transformed representation of the asymptotic wave functions can easily be verified for potential scattering. The asymptotic radial wave function in a given ℓ-channel satisfies the identity

$$k^{-\frac{1}{2}} e^{i\eta} \sin\left(kr - \frac{1}{2}\ell\pi + \eta\right) = k^{-\frac{1}{2}} \left[\sin\left(kr - \frac{1}{2}\ell\pi\right) + e^{i(kr - \frac{1}{2}\ell\pi)} T\right],$$

where $T = (e^{2i\eta} - 1)/2i = e^{i\eta} \sin\eta$.

Using these asymptotic continuum functions, the derivation given above implies that for $T_t = -m_{11}^{-1} m_{10}$, such that $m_{10} + m_{11} T_t = 0$, then

$$[T] = T_t - 2\Xi(T_t) = -2\left(m_{00} - m^{\dagger}_{10} m_{11}^{-1} m_{10}\right)$$

is stationary for infinitesimal variations of T_t, and for infinitesimal variations about an exact solution for which $\Xi = 0$. The significance of this revised formalism is that the matrix m_{11}, constructed from complex exponential basis functions, is a symmetric matrix but no longer real. Its null values are displaced away from the real energy axis, which essentially eliminates the problem of Kohn anomalies [237]. This can be rationalized from the theory of scattering resonances: null values of m_{11} correspond to pure outgoing-wave solutions of the projected Schrödinger equation. This condition characterizes resonances that in general are displaced below the real energy-axis by a finite width or reciprocal lifetime [192, 382].

Introduced in the context of heavy-particle reactive collisions [440], the complex Kohn method has been successfully applied to electron–molecule scattering [341]. It is accurate but computationally intensive, since continuum basis orbitals do not have the Gaussian form that is exploited in most *ab initio* molecular bound-state studies. The method has been implemented using special numerical methods [341] developed for these integrals. These numerical methods mitigate another practical limitation of the matrix variational method [270], in that analytic formulas are available only for specialized forms of the long-range potentials that can occur, especially for molecules.

8.3 Schwinger variational theory

In the Hulthén–Kohn formalism of the matrix variational method, a large part of the computational effort is concerned with the evaluation and inversion of the bound–bound matrix of the $(N+1)$-electron operator $H - E$. Because powerful computational methods for this matrix can be taken over directly from bound-state atomic physics and theoretical chemistry, there is a great practical advantage in using continuum variational methods that incorporate this as a central aspect of the overall algorithm. A fundamentally different approach replaces the Schrödinger equation by a formally equivalent Lippmann–Schwinger integral equation [228], developing stationary expressions for the resulting scattering matrices. In such methods, a Green function is introduced for the asymptotic electron continuum part of the overall wave function, while short-range effects are treated through an incremental potential Δv. This formalism makes use of the Schwinger variational principle [371]. Formulated as a variational method using exponential or Gaussian orbital basis functions [419, 418], and extended to a multichannel formalism [270, 396], it has been widely applied to calculations of electron–molecule scattering [230, 177].

The theory will be developed here with reference to a spherical coordinate system, appropriate to a target atom or to a single-center representation of a molecular wave function. Extension to a multicenter representation is straightforward, following the same formal argument, but greatly complicates the notation and the detailed form of the Green function used in this theory. The formalism can most easily be understood in the simple model of potential scattering in a single ℓ-channel. A model Schrödinger equation, determined by a radial Hamiltonian operator $h = \hat{t} + v(r)$, is assumed to be modified by a difference potential Δv, which may be a nonlocal linear operator. At energy $\epsilon = \frac{1}{2}k^2$, in the scattering continuum, regular and irregular solutions $w_0(r)$ and $w_1(r)$, respectively, of the model equation

$$(h - \epsilon)w_i(r) = 0$$

are assumed to be known. If normalized to unit Wronskian,

$$w_1(r)w_0'(r) - w_1'(r)w_0(r) = 1,$$

these functions define a Green function G, a linear operator whose kernel is

$$g(r, r') = 2w_0(r_<)w_1(r_>).$$

Acting on any function $F(r)$ that vanishes for $r \to \infty$ and is regular at the origin $r \to 0$,

$$GF(r) = 2\left[w_0(r)\int_r^\infty w_1(r') + w_1(r)\int_0^r w_0(r')\right]F(r')\,dr'.$$

This implies that $(h-\epsilon)GF = F$, which can be verified by applying the operator $h-\epsilon$ to $GF(r)$. Thus the Green function is a formal inverse of $h-\epsilon$. The kernel function $g(r,r')$ is symmetric in r, r', regular at the origin in either variable, and continuous. The derivative in either variable is discontinuous at $r'=r$, so that $(h-\epsilon)G$ is equivalent to a Dirac δ-function. G depends on ϵ by construction.

$$GF \sim 2w_0(r)\int_0^\infty w_1(r')F(r')\,dr'$$

at the coordinate origin, $r \to 0$, and as $r \to \infty$,

$$GF \sim 2w_1(r)\int_0^\infty w_0(r')F(r')\,dr'.$$

Thus GF is regular at the origin but is asymptotically proportional to the irregular function $w_1(r)$. It has the properties assumed for the second continuum basis function required for each open channel in the matrix variational method.

These properties of the model Green function imply that the Lippmann–Schwinger equation [228],

$$f = w_0 - G\Delta v f,$$

defines a solution of the radial Schrödinger equation $(h-\epsilon+\Delta v)f(r) = 0$. Regular and irregular functions with asymptotic forms, respectively,

$$w_0 \sim k^{-\frac{1}{2}}\sin\left(kr - \frac{1}{2}\ell\pi\right)a, \qquad w_1 \sim k^{-\frac{1}{2}}\cos\left(kr - \frac{1}{2}\ell\pi\right),$$

are used to construct the *principal value* Green function. This Green function imposes the asymptotic form $f(r) \sim w_0(r) + w_1(r)\tan\eta$, where

$$\tan\eta = -2\int_0^\infty w_0(r')\Delta v(r')f(r')\,dr'.$$

Other asymptotic forms consistent with unit Wronskian define different but equally valid Green functions, with different values of the asymptotic coefficient of w_1. In particular, if $w_1 \sim k^{-\frac{1}{2}}\exp i(kr - \frac{1}{2}\ell\pi)$, this determines the *outgoing-wave* Green function, and the asymptotic coefficient of w_1 is the single-channel T-matrix, $e^{i\eta}\sin\eta$. This is the basis of the *T-matrix* method [342, 344], which has been used for electron–molecule scattering calculations [126]. It is assumed that $\Delta v f$ is regular at the origin and that Δv vanishes more rapidly than r^{-2} for $r \to \infty$.

Coulomb or static dipole potentials should be included in the model potential whose solutions w_0, w_1 are used to construct the Green function.

Specializing the present derivation to the principal value Green function, the unsymmetrical expression $\tan\eta = -2(w_0|\Delta v|f)$ is exact for an exact solution of the Lippmann–Schwinger equation, but it is not stationary with respect to infinitesimal variations about such a solution. Since $w_0 = f + G\Delta v f$ for such a solution, this can be substituted into the unsymmetrical formula to give an alternative, symmetrical expression $\tan\eta = -2(f|\Delta v + \Delta v G \Delta v|f)$, which is also not stationary. However, these expressions can be combined to define the Schwinger functional

$$[\tan\eta] = -2(w_0|\Delta v|f)(f|\Delta v + \Delta v G \Delta v|f)^{-1}(f|\Delta v|w_0),$$

which is stationary for variations of the trial function f about an exact solution. The variation $\delta[\tan\eta]$ induced by an infinitesimal variation of f about an exact solution is

$$\left[\frac{(\delta f|\Delta v|w_0)}{(f|\Delta v|w_0)} - \frac{(\delta f|\Delta v + \Delta v G \Delta v|f)}{(f|\Delta v + \Delta v G \Delta v|f)} + hc\right][\tan\eta]$$
$$= -2(\delta f\,\Delta v|w_0 - f - G\Delta v f) + hc.$$

This implies that $[\tan\eta]$ is stationary if and only if f, in the range of Δv, satisfies the Lippmann–Schwinger equation.

The Schwinger functional has several remarkable properties. It combines three formulas that may have different numerical values:

$$\tan\eta \simeq -2(w_0|\Delta v|f),$$
$$\tan\eta \simeq -2(f|\Delta v + \Delta v G \Delta v|f),$$
$$\tan\eta \simeq -2(f|\Delta v|w_0).$$

The function f can be arbitrarily normalized, because $[\tan\eta]$ is homogeneous in f. There is no constraint on the asymptotic form of trial functions, because they have no effect outside the range of the assumed short-range potential Δv. Only the regular model functions w_0 occur explicitly in the theory, although the irregular functions w_1 must be consistent with the asymptotic character of the Green function, and may be used to construct it.

An algebraic theory is obtained if f is expanded within the range of Δv as a linear combination $f(r) = \sum_a \eta_a(r)(a|f)$ of orthonormal basis functions $\{\eta_a\}$. Δv is expanded in the same basis as a linear integral operator with the kernel,

$$\Delta v(r, r') = \sum_{a,b} \eta_a(r)\Delta v_{ab}\eta_b^*(r').$$

For $[\tan\eta]$ to be stationary with respect to variations of the coefficients $(a|f)$, this requires, for all indices a,

$$(a|f - w_0 + G\Delta v f) = 0.$$

This is a system of linear equations that determine

$$(a|f) = \sum_b (I + G\Delta v)^{-1}_{ab}(b|w_0)$$
$$= \sum_{b,c} (\Delta v + \Delta v G \Delta v)^{-1}_{ab} \Delta v_{bc}(c|w_0).$$

Combining these equations, all three expressions for $\tan\eta$ give the same result for the stationary functional,

$$[\tan\eta] = \sum_{a,b}\sum_{c,d}(w_0|a)\Delta v_{ac}(\Delta v + \Delta v G \Delta v)^{-1}_{cd}\Delta v_{db}(b|w_0).$$

8.3.1 Multichannel Schwinger theory

For practical use in electron scattering theory, the Lippmann–Schwinger equation and the Schwinger variational principle must be generalized to a multichannel formalism. In applying this formalism, the orthogonality conditions that distinguish the bound and free components of an $(N+1)$-electron continuum Ψ_s, as given in Eq. (8.1), must be taken into account. The argument here summarizes an earlier derivation [270]. The multichannel close-coupling equations appropriate to the $(N+1)$-electron problem require orthogonality constraints expressed by off-diagonal Lagrange multipliers. Reduction to a form in which these parameters vanish by construction will be considered here after generalizing the Schwinger stationary principle to such equations in their simpler homogeneous form.

Consider the multichannel radial close-coupling equations, without Lagrange multipliers for orthogonality,

$$\sum_q m^{pq} u_{qs}(r) = 0, \tag{8.11}$$

where

$$m^{pq} = (h - \epsilon)^{pq} + \Delta v^{pq}.$$

It is assumed that all closed-channel components have been eliminated by the partitioning transformations that determine Eq. (8.4), from which the radial operator m^{pq} is derived. Equations of this form can be solved using the Green function of the model problem

$$\sum_s (h - \epsilon)^{ps} w_{isq}(r) = 0.$$

The multichannel model functions w_{0sq} and w_{1sq} are real-valued continuum solutions of the model problem, in a K-matrix formalism. They are normalized to the matrix Wronskian condition

$$\sum_s (w'_{isp} w_{jsq} - w_{isp} w'_{jsq}) = \delta_{pq}(\delta_{i0}\delta_{j1} - \delta_{i1}\delta_{j0}).$$

The equations here are simplified by working in an eigenchannel representation, in which the model K-matrix is diagonalized, with eigenvalues $\tan\eta_\sigma$. Defining matrices C and S, diagonal in the eigenchannel representation, with eigenvalues $\cos\eta_\sigma$ and $\sin\eta_\sigma$, respectively, the physical K-matrix implied by an incremental matrix K' in this representation is [375]

$$K = (S + CK')(C - SK')^{-1}.$$

The multichannel generalization of the principal value Green function is a linear operator G^σ_{pq}, defined by its kernel

$$g^\sigma_{pq}(r, r') = 2w_{0p\sigma}(r_<)w_{1q\sigma}(r_>),$$

indexed by eigenchannel σ of the model continuum functions.

If $\{F_q(r)\}$ is a vector of functions, one for each open channel, that are regular at the coordinate origin and quadratically integrable, then

$$\sum_q G^\sigma_{pq} F_q(r) = \sum_q 2\left\{w_{0p\sigma}(r)\int_r^\infty w_{1q\sigma}(r') + w_{1p\sigma}(r)\int_0^r w_{0q\sigma}(r')\right\} F_q(r')\,dr',$$

such that

$$\sum_\sigma \sum_{p,q'} (h - \epsilon)^{pp'} G^\sigma_{p'q} F_q(r) = F_p(r).$$

Using the matrix Wronskian condition, the proof is the same as for the single-channel problem.

This multichannel matrix Green function determines a multichannel Lippmann–Schwinger equation

$$u_{p\tau} = w_{0p\tau} - \sum_\sigma \sum_{p',q} G^\sigma_{pp'} \Delta v^{p'q} u_{q\tau}$$

that determines solutions of the multichannel Schrödinger equation. The asymptotic form of the second term here is

$$-2\sum_{p,q} w_{1p\sigma}(w_{0p\sigma}|\Delta v^{pq}|u_{q\tau}),$$

which implies that the incremental K-matrix in the model eigenchannel representation is $K' = -2(w_0|\Delta v|u)$, in a matrix notation. The Hermitian conjugate relation is $K' = -2(u|\Delta v|w_0)$.

The multichannel Lippmann–Schwinger equation implies that $w_0 = u + G\Delta v u$. Substituting this into the second expression for K' gives an alternative formula,

$$K' = -2(u|\Delta v + \Delta v G \Delta v|u).$$

The multichannel Schwinger functional is defined by the matrix product

$$[K'] = -2(w_0|\Delta v|u)(u|\Delta v + \Delta v G \Delta v|u)^{-1}(u|\Delta v|w_0).$$

This expression is stationary for variations of u if and only if u is an exact solution of the multichannel Lippmann–Schwinger equation. The proof follows exactly as in the single-channel case, if the order of matrix products is maintained in the derivation. If the multichannel continuum solution is expanded as

$$u_{ps}(r) = \sum_a \eta_{pa}(pa|ps),$$

where $\{\eta_{pa}\}$ are orthonormal radial basis functions, then $[K']$ is a function of the coefficients $(pa|ps)$. Following the logic of the single-channel derivation, the stationary expression obtained by varying $[K']$ with respect to these coefficients is

$$[K']_{\sigma\tau} = -2\sum_{a,b}\sum_{p,p'}\sum_{q,q'}(w_{0p\sigma}|\Delta v|\eta_{p'a})[(\eta|\Delta v + \Delta v G \Delta v|\eta)]^{-1}_{p'a,q'b}(\eta_{q'b}|\Delta v|w_{0q\tau}).$$

The same result is obtained by substituting a formal solution for u into any of the three approximate expressions for K' combined in the Schwinger functional matrix.

8.3.2 Orthogonalization and transfer invariance

Equations (8.11), the multichannel radial close-coupling equations, are derived from Eq. (8.3) by projecting onto orbital spherical harmonics for a single spin component. The partitioning argument that leads to Eq. (8.3) explicitly requires each channel orbital wave function ψ_{ps} to be orthogonal to all orbital functions used to construct target states Θ_p and $(N+1)$-electron basis states Φ_μ. In the matrix variational formalism, this condition is imposed by orthogonalizing each continuum basis function to the full set of quadratically integrable basis orbitals. Because open-channel continuum functions have finite amplitudes outside any sphere enclosing the target system, such orthogonalization cannot eliminate them. In practice, this is also true for pseudostate closed-channel orbitals generated by long-range potentials.

This orthogonality condition is implemented in close-coupling theory by introducing Lagrange multipliers λ^p_{as} into the radial integrodifferential equations, which become

$$\sum_q m^{pq} u_{qs}(r) = \sum_a \eta_{pa} \lambda^p_{as}. \tag{8.12}$$

The λ multipliers are determined by the orthogonality condition. If the quadratically integrable radial basis in channel p is the orthonormal set $\{\eta_{pa}\}$, values of the Lagrange multipliers are given by

$$\lambda^p_{as} = \sum_q (\eta_{pa}|m^{pq}|u_{ps}).$$

In the Lippmann–Schwinger formalism, the Green function can be defined in a projected orbital space consistent with this orthogonality condition. An alternative, assumed in the derivation given above, is to use a Green function appropriate to close-coupling equations from which the Lagrange multipliers have been eliminated. This can be done if each $(N+1)$-electron trial wave function has the property of *transfer invariance* [270]. This means that Ψ_s, in Eq. (8.1), is unchanged under any transfer of quadratically integrable terms $\mathcal{A}\Theta_p\phi_{ps}$ between its free and bound components, $\sum_p \mathcal{A}\Theta_p\psi_{ps}$ and $\sum_\mu \Phi_\mu c_{\mu s}$, respectively. If a valid system of close-coupling equations exists for any arbitrary transfer of such terms, these equations are *transfer covariant*. Projection of the Schrödinger equation onto the bound space determines the coefficients $c_{\mu s}$ through linear equations

$$(\Phi_\mu|H - E|\Psi_s) = 0$$

for all μ, or

$$\sum_\nu (H - E)_{\mu\nu} c_{\nu s} = -\sum_p (\Phi_\mu|H - E|\mathcal{A}\Theta_p\psi_{ps}).$$

This is transfer covariant if all quadratically integrable functions are represented in the same orbital basis. Requiring ψ_{ps} to be orthogonal to all ϕ_{pa} (radial factor $\eta_{pa}(r)$) enforces a unique representation, but introduces Lagrange multipliers in the close-coupling equations. An alternative is to require

$$(\Theta_p|H - E|\Psi_s) = 0$$

to be valid for all Θ_p, and to adjust the Lagrange multipliers accordingly. This implies that

$$(\mathcal{A}\Theta_p\phi_{pa}|H - E|\Psi_s) = 0$$

for all p, a. This reduces to

$$\sum_q (\eta_{pa}|m^{pq}|u_{qs}) = \lambda^p_{as} = 0,$$

which removes the Lagrange multipliers, if the operator m^{pq} is determined from $(\Theta_p|H - E|\Psi_s)$ as in the matrix variational formalism. This analysis shows that a representation of the transfer-covariant close-coupling equations exists in which the Lagrange multipliers for orthogonalization vanish. This representation is assumed in the development of the multichannel Schwinger theory given above.

Use of the model Green function in the Lippmann–Schwinger equation is the most serious weakness of the Schwinger variational formalism. The method requires Δv to be a short-range potential or operator, so that it can be represented quantitatively in a quadratically integrable orbital basis. This requires all long-range potentials to be included in the model Hamiltonian used to construct the Green function. This is a formidable difficulty in the context of electron–molecule scattering, since much of the structure of low-energy cross sections is due to polarization response and to long-range dipole and quadrupole potentials [215, 178]. A molecular Green function must satisfy Coulomb cusp conditions and be regular at each atomic center, precluding the use of standard asymptotic forms for model long-range potentials. Construction of such a Green function and evaluation of the implied integrals $(\eta|\Delta v G \Delta v|\eta)$ requires analytic or computational methodology beyond present capabilities.

8.4 Variational *R*-matrix theory

The Wigner–Eisenbud [428] R matrix, or derivative matrix, is defined by the relationship between radial channel orbitals $f_{ps}(r)$ and their derivatives on some sphere of radius r_1 that surrounds a target system [214, 33]. Assuming spherical geometry, the dimensionless radial R-matrix R_{pq} at r_1 is defined by

$$f_{ps}(r_1) = \sum_q R_{pq} r_1 f'_{qs}(r_1).$$

The indices refer to both open and closed channels, which do not have to be distinguished at the finite radius r_1. As defined for solutions of radial close-coupling equations, R_{pq} is a real symmetric matrix. Extension of the theory and definitions to more general geometry will be considered below. For electron–molecule scattering, a molecular center is chosen somewhat arbitrarily such that the electronic charge is approximately enclosed in some smaller sphere of radius r_0, while bound–free exchange can be neglected outside the larger R-matrix radius r_1.

The theory of the R-matrix was developed in nuclear physics. As usually presented, the theory makes use of a Green function to relate value and slope of the radial channel orbitals at r_1, expanding these functions for $r < r_1$ as linear combinations of basis functions that satisfy fixed boundary conditions at r_1. The true logarithmic derivative (or reciprocal of the R-matrix in multichannel formalism) is computed from Green's theorem, despite the use of basis functions whose logarithmic derivatives at r_1 have a fixed but arbitrary value. Because of the inherent discontinuity of the boundary derivative this expansion tends to converge slowly, and requires correction by an approximate method due to Buttle [54].

In nuclear physics, the specifically nuclear interaction is of short range, so that full scattering information is obtained by matching the R-matrix at r_1 to asymptotic external wave functions. The method has been extended to electron–atom scattering, where long-range potentials are important, by combining basis expansion of the channel orbital functions within r_1 with explicit numerical solution of close-coupling equations outside [50, 47]. This method makes it possible to process algebraic equations defined by matrix elements of nonlocal operators within r_1, while solving the simple asymptotic close-coupling equations, without exchange, outside r_1. This requires r_1 to be large enough that exchange can be neglected and that the nonlocal optical potential can be approximated by an asymptotic local potential.

The R-matrix can be matched at r_1 to external channel orbitals, solutions in principle of external close-coupling equations, to determine scattering matrices. Radial channel orbital vectors, of standard asymptotic form for the K-matrix,

$$w_{0pq}(r) \sim k_p^{-\frac{1}{2}} \sin\left(k_p r - \frac{1}{2}\ell_p \pi\right) \delta_{pq},$$

$$w_{1pq}(r) \sim k_p^{-\frac{1}{2}} \cos\left(k_p r - \frac{1}{2}\ell_p \pi\right) \delta_{pq},$$

are defined by integrating inwards for $r > r_1$. The asymptotic forms must be modified appropriately for Coulomb or fixed-dipole scattering. The function to be fitted at r_1 is

$$f_{ps}(r_1) = \sum_q [w_{0pq}(r_1)\delta_{qs} + w_{1pq}(r_1)K_{qs}].$$

The R-matrix relation between function value and gradient can be solved for the K-matrix,

$$K_{st} = -\sum_p \left[w_{1ps}(r_1) - \sum_q r_1 R_{pq}(r_1) w'_{1qs}(r_1) \right]^{-1}_{sp} \times \left[w_{0pt}(r_1) - \sum_q r_1 R_{pq}(r_1) w'_{0qt}(r_1) \right]_{pt}.$$

As written, these equations refer to open channels only. When external closed channels are considered, an external closed channel orbital that vanishes as $r \to \infty$ must be included for each such channel. The indices p, q, s run over all channels, but the index t refers to open channels only. K_{st} is a rectangular matrix, whose open-channel submatrix can be shown to be real and symmetric.

The theory of the R-matrix can be understood most clearly in a variational formulation. The essential derivation for a single channel was given by Kohn [202], as a variational principle for the radial logarithmic derivative. If h is the radial Hamiltonian operator, the Schrödinger variational functional is

$$\Xi = \int_0^\infty f(h-\epsilon)f\,dr.$$

If $f(r)$ is an exact solution for $r > r_1$, this reduces the functional to the finite integral considered by Kohn,

$$\Xi_\lambda = \int_0^{r_1} f(h-\epsilon)f\,dr = \int_0^{r_1}\left[\frac{1}{2}(f')^2 + \left(v(r) - \frac{1}{2}k^2\right)f^2\right]dr - \frac{1}{2}\lambda f^2(r_1),$$

obtained after integration by parts. The parameter λ is the logarithmic derivative $f'/f = (r_1 R)^{-1}$, defining the parameter R as the single-channel R-matrix.

For an infinitesimal variation δf, with fixed λ, the variation of Ξ_λ is

$$\delta\Xi_\lambda = 2\int_0^{r_1} \delta f(h-\epsilon)f\,dr + \delta f(r_1)[f'(r_1) - \lambda f(r_1)].$$

This vanishes for unconstrained variations if and only if $(h-\epsilon)f = 0$ for $0 \leq r \leq r_1$ and $\lambda = f'(r_1)/f(r_1)$.

If $f(r)$ is approximated by a finite expansion in linearly independent basis functions $\{\eta_a\}$, such that $f = \sum_a \eta_a c_a$, variation of Ξ_λ with respect to the coefficients c_a gives the linear equation

$$\sum_b A_{ab}c_b = \frac{1}{2}\lambda\eta_a(r_1)f(r_1), \tag{8.13}$$

where

$$A_{ab} = \int_0^{r_1}\left[\frac{1}{2}\eta'_a\eta'_b + \eta_a\left(v(r) - \frac{1}{2}k^2\right)\eta_b\right]dr. \tag{8.14}$$

The parameter λ is uniquely determined as a consistency condition [202]. From Eq. (8.13),

$$c_a = \frac{1}{2}\lambda\sum_b (A^{-1})_{ab}\eta_b(r_1)f(r_1),$$

so that

$$f(r_1) = \sum_a \eta_a(r_1)c_a = \frac{1}{2}\lambda\sum_{a,b}\eta_a(r_1)(A^{-1})_{ab}\eta_b(r_1)f(r_1).$$

Unless $f(r_1)$ vanishes, this requires as a consistency condition

$$R = (r_1\lambda)^{-1} = \frac{1}{2r_1}\sum_{a,b}\eta_a(r_1)(A^{-1})_{ab}\eta_b(r_1),$$

which is the standard formula for the single-channel R-matrix.

Several important aspects of this formula generalize immediately to multichannel scattering. The matrix A_{ab} is the Hermitian residue of Ξ after integration by parts, which separates off the boundary term as a surface integral. This boundary term determines the R-matrix, which contains all information needed to extend an internal solution by matching to external wave functions. Thus the internal Schrödinger equation, which may contain nonlocal potentials such as an optical potential, is solved by the typical bound-state method of expanding in a set of basis functions, while the external problem may be solved by some completely independent method. If r_1 is large enough that asymptotic potentials are valid for $r > r_1$, then specialized and highly efficient analytic methods may be used, without affecting the variational expansion appropriate to the inner region.

A multichannel generalization of the variational R-matrix theory was derived by Jackson [183]. In the context of the generalized close-coupling equations, incorporating an optical potential derived by Feshbach partitioning, this is in principle an exact theory. Exact solutions of the coupled equations are assumed for $r \geq r_1$, determined by specifying their values at r_1 in terms of an R-matrix. For each channel p, projected onto spin and spatial symmetry (angular) functions, there are two linearly independent radial functions. If these are denoted by $u_{0ps}(r)$ and $u_{1ps}(r)$, respectively, for a global wave function indexed by s, the boundary conditions at r_1 chosen by Jackson are

$$u_{0ps}(r_1) = 0, \qquad u'_{0ps}(r_1) = \delta_{ps} r_1^{-\frac{1}{2}}$$

$$u_{1ps}(r_1) = \delta_{ps} r_1^{+\frac{1}{2}}, \qquad u'_{1ps}(r_1) = 0.$$

In this R-matrix theory, open and closed channels are not distinguished, but the eventual transformation to a K-matrix requires setting the coefficients of exponentially increasing closed-channel functions to zero. Since the channel functions satisfy the unit matrix Wronskian condition, a generalized Kohn variational principle is established [195], as in the complex Kohn theory. In this case the canonical form of the multichannel coefficient matrices is

$$\alpha_0 = I, \qquad \alpha_1 = R$$

and the radial channel orbital function is

$$f_{ps}(r) = \sum_t (u_{0pt}\delta_{ts} + u_{1pt}R_{ts}), \qquad r \geq r_1.$$

Then, by construction,

$$f_{ps}(r_1) = R_{ps} r_1^{\frac{1}{2}}, \qquad f'_{ps}(r_1) = \delta_{ps} r_1^{-\frac{1}{2}},$$

in agreement with the defining equation for the R-matrix,

$$f_{ps}(r_1) = \sum_q R_{pq} r_1 f'_{qs}(r_1).$$

Since an exact solution is assumed outside r_1, the Schrödinger matrix functional is

$$\Xi_{st} = \sum_{p,q} \int_0^{r_1} f^{\dagger}_{ps}(r) m^{pq} f_{qt}(r)\, dr,$$

whose variation is given by Eq. (8.9). This defines a trial R-matrix $R_t = -m_{11}^{-1} m_{10}$, such that $m_{10} + m_{11} R_t = 0$ in this representation of the channel orbitals, and implies that $[R] = R_t - 2\Xi(R_t)$ is stationary. This defines

$$[R] = -2\left(m_{00} - m^{\dagger}_{10} m_{11}^{-1} m_{10}\right). \qquad (8.15)$$

This expression is stationary with respect to variations about an exact solution.

Because radial derivatives of the functions u_{1ps} all vanish at $r = r_1$, null values of the determinant of m_{11} characterize eigensolutions of the multichannel Schrödinger equation that satisfy a homogeneous Neumann boundary condition on the R-matrix sphere. In contrast to the Kohn formalism for the K-matrix, these null values are not anomalous, but are a characteristic feature of any system of integrodifferential equations as energy or other parameters are varied. In one dimension, this simply means that the logarithmic derivative passes through zero, as a continuous nondecreasing function of energy.

The derivation by Kohn [202] can readily be extended to the multichannel R-matrix. The underlying logic depends on the variational principle that the multichannel Schrödinger functional is stationary for variations of a trial function that satisfies the correct boundary conditions if and only if that function satisfies the Schrödinger equation. In a matrix notation, suppressing summations and indices, the variational functional of Schrödinger is

$$\Xi = \int f^{\dagger} \hat{A} f - \frac{1}{2r_1} f^{\dagger}(r_1) R^{-1} f(r_1),$$

when integrated by parts to separate the surface term from the residual Hermitian volume term $\int f^{\dagger} \hat{A} f$. Here R is a trial R-matrix, analogous to a Lagrange multiplier, to be adjusted so that the trial wave function satisfies R-matrix boundary conditions at $r = r_1$. $\hat{A}$ defines the modified Hamiltonian operator of Bloch [30], such that

$$h_B - \epsilon = h - \epsilon + \frac{1}{2}\delta(r - r_1)\frac{d}{dr}.$$

Its representation in an orbital basis $\{\eta_a\}$ defines the matrix A given by Eq. (8.14). Given $f = \eta c$, the functional becomes

$$\Xi = c^\dagger \left[A - \frac{1}{2r_1}\eta^\dagger(r_1)R^{-1}\eta(r_1) \right] c.$$

Variation with respect to the coefficients $c^\dagger$, for fixed R, gives the linear equations

$$Ac = \frac{1}{2r_1}\eta^\dagger(r_1)R^{-1}f(r_1),$$

whose solution is

$$c = \frac{1}{2r_1}A^{-1}\eta^\dagger(r_1)R^{-1}f(r_1).$$

This implies that $\Xi = 0$ and

$$f(r_1) = \eta(r_1)c = \frac{1}{2r_1}\eta(r_1)A^{-1}\eta^\dagger(r_1)R^{-1}f(r_1).$$

Unless A is singular, as it is at the null values of m_{11}, this implies the consistency condition, restoring summation indices,

$$R_{pq} = \frac{1}{2r_1}\sum_{a,b}\eta_a(r_1)(A^{-1})^{pq}_{ab}\eta^\dagger_b(r_1).$$

Since matrix A is Hermitian by construction, the R-matrix is also Hermitian, and can be made real by suitable choice of phase factors for the basis functions. Because Ξ vanishes, R_{pq} is the stationary value of the matrix $[R]$ in the basis $\{\eta_a\}$.

These early derivations [202, 183] have an important feature, not emphasized at the time, that has subsequently turned out to be of great practical significance. No conditions are imposed on the basis functions η other than regularity at the coordinate origin and linear independence. The nonvariational theory [428], which uses Green's theorem, introduces a complete basis set in the interval $0 \le r < r_1$, obtained as the set of eigenfunctions of some model Hamiltonian, with a fixed boundary condition, $r_1\eta'(r_1) = b\eta(r_1)$. Several standard derivations follow this logic [214, 47]. If the trial function $f(r)$ is expanded as a sum of functions all with the same logarithmic derivative at r_1, it must satisfy this same boundary condition. The value of λ computed from the variational formula in the Kohn theory differs in general from the imposed value b/r_1. Unless the energy parameter is an exact eigenvalue of the A-matrix, with this imposed boundary condition, the trial function in standard R-matrix theory has a discontinuous derivative at r_1, if the function value is matched at the boundary. This discontinuity leads to slow convergence of the basis set expansion [187]. Standard procedure is to remove the gradient discontinuity by a nonvariational Buttle correction [54]. This methodology has been used in an extended series of *ab initio* calculations of atomic [50, 47] and molecular [138, 139, 249] scattering cross sections. An alternative method [106, 220] is to adjust the parameter b iteratively so that it agrees with the computed value of λ. Other calculations have used the variational formalism directly, implemented with orbital basis functions that are not constrained at r_1 [310, 311, 299, 298].

The R-matrix radius r_1 must be large enough that all nonlocal interactions can be neglected or replaced by local asymptotic potentials outside it. In electron–molecule scattering, a much smaller target radius r_0 can be defined, within which the total target electronic density has converged sufficiently to represent molecular multipole moments and polarizabilities within the overall accuracy of the full scattering calculation. In the intermediate region $r_0 \leq r \leq r_1$, basis orbital functions for the target system die off exponentially, while external multipole and polarization response potentials are well approximated by their asymptotic forms. This behavior has several very important implications for quantitative *ab initio* calculations. Crucially, orbital basis sets and multiconfiguration expansions adapted to the target system cannot adequately represent the intermediate region. Open-channel orbital functions retain a constant amplitude throughout this region, and closed-channel (polarization pseudostate) orbitals decay as inverse powers of r, not exponentially. The most direct way to represent this region is to add orthogonalized fixed-energy continuum orbital functions, two for each open channel, to the basis set. When such basis orbitals are orthogonalized to the exponential basis set, this induces an orthogonal projection of the $(N+1)$-electron variational basis, as indicated in the definition of the $(N+1)$-electron wave function, Eq. (8.1). Fixed linear combinations of Gaussian basis orbitals (with small exponents) can be constructed that model spherical Bessel functions or Coulomb functions quite accurately in the intermediate region [300]. An alternative basis for the intermediate region is the set of *numerical asymptotic functions* (NAFs), continuum eigenfunctions of the model Hamiltonian constructed from the long-range asymptotic potential functions, integrated outwards from r_0 [311]. Since the centrifugal potential barrier increases rapidly with orbital angular momentum ℓ, orbital functions with high ℓ values become insensitive to the region inside r_0. This justifies the *asymptotic distorted-wave* (ADW) approximation [273], in which the R-matrix is computed by numerical integration of the asymptotic close-coupling equations from r_0 to r_1. The high-ℓ channel sectors of this matrix are used to augment a variational R-matrix computed in an orbital basis cut off at some relatively smaller ℓ-value. NAF orbitals and the ADW approximation were used in accurate calculations of fixed-nuclei e^-–H_2 scattering cross sections [299, 298].

8.4.1 Variational theory of the $\mathcal{R}$-operator

Let τ denote a spatial cell of finite volume, enclosed by a convex bounding surface σ. Variational R-matrix theory can be put into the general context of the classical theory of elliptical partial differential equations by considering the three-dimensional Schrödinger equation (for a given spin index) within such a bounding surface [272]. For a scalar wave function $\psi(\mathbf{x})$ with an implicit spin index, the Schrödinger functional is an integral over the volume τ

$$\Xi = \int_\tau \psi^*(h-\epsilon)\psi = A - \frac{1}{2}\int_\sigma \psi^*\nabla_n\psi,$$

integrating the kinetic energy term by parts. $\nabla_n\psi$ here is the outward normal gradient on σ of the trial function ψ. The residual volume integral A, incorporating a Bloch-modified [30] kinetic energy term, is

$$A = \int_\tau \left[\frac{1}{2}\nabla\psi^* \cdot \nabla\psi + \psi^*(v-\epsilon)\psi\right].$$

In any representation basis $\{\phi_a(\mathbf{x})\}$, the matrix

$$A_{ab} = \int_\tau \left[\frac{1}{2}\nabla\phi_a^* \cdot \nabla\phi_b + \phi_a^*(v-\epsilon)\phi_b\right]$$

is Hermitian if the potential v is real (or nonlocal and Hermitian).

If ψ and $\nabla_n\psi$ are given on the enclosing surface σ, the $\mathcal{R}$ operator is defined such that

$$\psi(\boldsymbol{\sigma}_1) = \int_\sigma \mathcal{R}(\boldsymbol{\sigma}_1, \boldsymbol{\sigma}_2)\nabla_n\psi(\boldsymbol{\sigma}_2)d^2\boldsymbol{\sigma}_2.$$

This will be symbolized here by $\psi(\sigma) = \mathcal{R}\nabla_n\psi(\sigma) = \mathcal{R}\xi(\sigma)$. If the normal gradient ξ is specified, this defines a classical Neumann boundary condition on σ, which determines a unique solution of the Schrödinger equation in the enclosed volume τ. The value of the boundary integral is

$$A_1 = \frac{1}{2}\int_\sigma \psi^*\xi.$$

The functional Ξ must vanish for an exact solution of the Schrödinger equation in τ. This implies that A_1 and A are equal, and because A is real,

$$A = A_1 = A_1^*.$$

Infinitesimal variations of ψ about an exact solution give

$$\delta A = \left(\int_\tau \delta\psi^*(h-\epsilon)\psi + \frac{1}{2}\int_\sigma \delta\psi^*\nabla_n\psi + cc\right),$$

and

$$\delta A_1 = \frac{1}{2}\int_\sigma \delta\psi^*\xi.$$

In analogy to the Schwinger variational principle, consider the product functional

$$[A] = A_1 A^{-1} A_1^*,$$

which is real for any trial function. From the variations of A_1 and A,

$$\delta[A] = [A]\left[A_1^{-1}\frac{1}{2}\int_\sigma \delta\psi^*\xi - A^{-1}\left(\frac{1}{2}\int_\sigma \delta\psi^*\nabla_n\psi + \int_\tau \delta\psi^*(h-\epsilon)\psi\right) + cc\right].$$

For variations about an exact solution, for which $A = A_1 = A_1^*$, this reduces to

$$\delta[A] = \left(\frac{1}{2}\int_\sigma \delta\psi^*(\xi - \nabla_n\psi) - \int_\tau \delta\psi^*(h-\epsilon)\psi + cc\right).$$

For variations of ψ that are unconstrained throughout τ and on σ, this implies that $[A]$ is stationary if and only if $(h-\epsilon)\psi = 0$ in τ and $\nabla_n\psi = \xi$ on σ.

When $\psi(\mathbf{x}) = \sum_a \phi_a(\mathbf{x})c_a$ is expressed as a sum of basis functions, variations $\delta\psi$ are driven by variations of the coefficients c_a. If $\delta[A]$ vanishes, this implies

$$\sum_b \int_\tau \phi_a^*(h-\epsilon)\phi_b c_b = \frac{1}{2}\int_\sigma \phi_a^*\left(\xi - \sum_b \nabla_n\phi_b c_b\right).$$

Integrating by parts, in terms of matrix elements A_{ab},

$$\sum_b A_{ab}c_b = \frac{1}{2}\int_\sigma \phi_a^*\xi, \quad \text{all } a.$$

Explicitly,

$$c_b = \frac{1}{2}\sum_a [A^{-1}]_{ba}\int_\sigma \phi_a^*\xi, \quad \text{all } b.$$

When these values are substituted into the definitions of A, A_1, and $[A]$, all of these quantities are equal to

$$[A] = \frac{1}{4}\sum_{a,b}\int_{\sigma_1}\int_{\sigma_2} \xi^*(1)\phi_a(1)[A^{-1}]_{ab}\phi^*(2)_b\xi(2).$$

In terms of the $\mathcal{R}$-operator, the definition of A_1 is

$$A_1 = \frac{1}{2}\int_{\sigma_1}\int_{\sigma_2} \xi^*(1)\mathcal{R}(1,2)\xi(2).$$

Comparison with the stationary value of $[A] = A_1$ indicates that

$$\mathcal{R}(1,2) = \frac{1}{2}\sum_{a,b} \phi_a(1)[A^{-1}]_{ab}\phi_b^*(2).$$

Because $[A]$ is stationary and the surface function ξ is arbitrary, the operator $\mathcal{R}$ must itself be stationary. It is evidently real and symmetric by construction. The basis functions $\phi_a(\mathbf{x})$ are required only to be linearly independent in τ. Enforcing orthonormality through a fixed boundary condition on σ imposes a lack of completeness with respect to determining the $\mathcal{R}$-operator. The factor $1/r_1$ is omitted from the definition given above, making the operator $\mathcal{R}$ dimensionally a length. This cannot be avoided in general geometry, since no unique radius is defined for a nonspherical surface. The usual R-matrix in spherical geometry is the surface-harmonic representation of the $\mathcal{R}$-operator as defined here, divided by the dimensional scale factor r_1.

8.4.2 The $\mathcal{R}$-operator in generalized geometry

In generalized geometry, it may be useful to introduce different coordinate systems on different sectors of a boundary hypersurface σ. This extended definition was introduced by Light and Walker [225] to define successive extensions of a propagated R-matrix, and implicitly by Schneider *et al.* [363] in extending R-matrix theory to vibronic interactions in molecules. It provides the formal basis for recent applications of adiabatic phase-matrix theory to rovibronic excitation in molecules [284]. It is assumed that coordinates can be defined such that the generalized kinetic energy operator in the enclosed hypervolume τ takes the form

$$\hat{T} = \frac{1}{2}\nabla\cdot\nabla,$$

for a generalized scalar product defined by

$$\mathbf{p}\cdot\mathbf{q} = \sum_i (1/\mu_i)p_i q_i.$$

The $\mathcal{R}$-operator is defined on hypersurface σ by

$$\psi(\sigma_1) = \int_\sigma \mathcal{R}(\sigma_1,\sigma_2)\boldsymbol{\nabla}\psi(\sigma_2)\cdot d\boldsymbol{\sigma}_2.$$

The classical Neumann boundary problem generalizes directly to the hyperspace if the kinetic energy operator can be put into this canonical form. The argument given above, when generalized to nonspherical geometry, remains valid. Given the normal gradient ξ on σ, the stationary value of the variational functional is

$$[A] = \frac{1}{2}\int_\sigma\int_\sigma d\sigma_1 \xi^*(\sigma_1)\mathcal{R}(\sigma_1,\sigma_2)\xi(\sigma_2)\,d\sigma_2.$$

Integration by parts of the Schrödinger functional is equivalent to using a Bloch-modified Schrödinger equation

$$(H_B - E)\psi(\mathbf{x}) = \int_\sigma \mathcal{L}(\mathbf{x},\boldsymbol{\sigma})\psi(\boldsymbol{\sigma})\,d\sigma.$$

Here the Bloch-modified Hamiltonian H_B is obtained by integration by parts of the kinetic energy integral,

$$\int_\tau \phi_\alpha^* \hat{T}\phi_\beta d\tau = \frac{1}{2}\int_\tau \nabla\phi_\alpha^* \cdot \nabla\phi_\beta d\tau - \frac{1}{2}\int_\sigma \phi_\alpha^* \nabla\phi_\beta \cdot d\boldsymbol{\sigma}.$$

This can be expressed in terms of a Bloch surface operator,

$$\mathcal{L} = \frac{1}{2}\delta(\mathbf{x},\boldsymbol{\sigma})\mathbf{n}(\boldsymbol{\sigma})\cdot\nabla$$

such that

$$\frac{1}{2}\int_\sigma \phi_\alpha^* \nabla\phi_\beta \cdot d\boldsymbol{\sigma} = \int_\tau \phi_\alpha^* \mathcal{L}\phi_\beta\, d\tau.$$

Expansion in a linear independent orbital basis implies

$$\mathcal{R}(\sigma_1,\sigma_2) = \frac{1}{2}\sum_\alpha\sum_\beta \phi_\alpha(\sigma_1)(H_B - E)^{-1}_{\alpha\beta}\phi_\beta^*(\sigma_2).$$

8.4.3 *Orbital functional theory of the R-matrix*

Variational R-matrix theory has been developed here in the context of multichannel close-coupling equations derived by Feshbach partitioning from the $(N+1)$-electron wave function given by Eq. (8.1). The short-range (exponentially decreasing) part of this wave function is eliminated in this formalism, transformed into a multichannel optical potential in the residual coupled orbital equations. Bound target states and pseudostates have been projected out to define the coupled open and closed "free" electronic channels. These equations take the form of a multichannel generalization of the ground-state orbital Euler–Lagrange equations of orbital functional theory, derived in Chapter 5 here. This relationship has been used to consider multipole polarization response as an example of the OFT formalism for electronic correlation [290].

Specializing to electronically elastic scattering, there is only one target reference state, which will be denoted by Φ, a normalized N-electron Slater determinant constructed from occupied OFT target orbitals. The target OEL equations are assumed to contain a correlation potential providing an accurate approximation to the true ground state. The $(N+1)$-electron reference state takes the form $\mathcal{A}\Phi\phi_\kappa$, where $\mathcal{A}$ is an antisymmetrizing operator. While all bound target orbital functions are normalized within the R-matrix boundary $r = r_1$, the continuum orbital ϕ_κ extends beyond this boundary. Its normalization requires some discussion.

Channel orbital functions are considered here to be orthogonalized to the bound-state orbital basis. If this basis is complete within the target radius r_0, these orthogonalized channel orbitals must effectively vanish inside the target system. In R-matrix formalism, if the effective potentials in the close-coupling equations do not depend on the multichannel orbital functions, these orbitals may be arbitrarily normalized [270]. Conventionally, they are normalized within the R-matrix volume characterized by radius r_1, where they are matched to logarithmic derivatives of external continuum wave functions normalized to unit flux density. This conventional normalization obscures the fact that a continuum orbital, extending throughout an infinite volume, must vanish in any finite volume if normalized to unity over all space. In the context of a continuum OFT, a continuum orbital may either be normalized to an infinitesimal within the R-matrix boundary or normalized to unity but assigned an infinitesimal occupation number. The latter option is followed here. In a set of N+1 occupied orbitals, the orbital wave functions are orthonormal for $r \leq r_1$, but one occupation number $n_\kappa \to 0+$. Thus $\sum_i n_i = N$, and bound–free exchange is described by a Fock operator that depends only on the occupied target orbitals. Similarly, in the theory developed here, correlation energy of the target is not affected by a continuum orbital, but bound–free correlation energy acts as an effective potential for the continuum electron.

In the case of a scattering resonance, bound–free correlation is modified by a transient "bound" state of N+1 electrons. In a finite matrix representation, the projected $(N+1)$-electron Hamiltonian $\hat{H}$ has positive energy eigenvalues, which define possible scattering resonances if they interact sufficiently weakly with the scattering continuum. In resonance theory [270], this transient discrete state is multiplied by an energy-dependent coefficient whose magnitude is determined by that of the channel orbital in the resonant channel. Thus the normalization of the channel orbital establishes the absolute amplitude of the transient discrete state, and arbitrary normalization of the channel orbital cannot lead to an inconsistency.

Bound–free correlation

For an N-electron target state, E_c is a sum of pair-correlation energies. In a two-electron system, such as atomic He, a major contribution to the correlation energy arises from replacement of a product function by one in relative coordinates that satisfies the Kato cusp condition at the singularity $r_{12} \to 0$. In a CI representation, this implies a slowly convergent sum of configurations, including relatively high angular quantum numbers. Methods used to simplify this problem include use of a free-electron-gas (FEG) local correlation potential [134] and an Ansatz wave function that specifically corrects the cusp condition [70]. Complementary to this short-range effect, the emphasis here is on long-range correlation potentials that arise from multipole response of the target to a scattered electron.

The N-electron target wave function is coupled to a continuum orbital ϕ_κ for which $n_\kappa \to 0$. Vanishing n_κ implies that the continuum electron does not modify the effective Hamiltonian $\mathcal{G}$ that acts on occupied target orbitals ($n_i = 1$). $\mathcal{G}$ also acts on ϕ_κ because $n_\kappa \to 0$ cancels out of the functional derivatives in $\frac{\delta E}{n_\kappa \delta \phi_\kappa^*}$. This implies that ϕ_κ is orthogonal to the occupied target orbitals. The result is to augment standard static-exchange equations with a nonlocal correlation potential $\hat{v}_c$.

From Eq. (5.7), and Janak's theorem [185], the contribution of correlation energy to the mean energy of the continuum orbital within the R-matrix boundary is

$$(\kappa|\hat{v}_c|\kappa) = \sum_j n_j \sum_{c<b} (1-n_c)(1-n_b)(\kappa j|\bar{u}|cb)(cb|\bar{c}|\kappa j)$$
$$- \sum_{k<j} n_k n_j \sum_b (1-n_b)(kj|\bar{u}|\kappa b)(\kappa b|\bar{c}|kj), \qquad (8.16)$$

where all integrals are evaluated for $r \le r_1$. The second line of this general formula is a correction to target correlation energy due to removal of ϕ_κ from the unoccupied set. The first line can be shown to include a multipole polarization potential.

To model the effect of close-coupling equations including a closed pseudostate channel, suppose that a pseudostate orbital ϕ_{p_j} represents the first-order perturbation of a particular target orbital ϕ_j in a multipole field of order λ, and that the pseudostate excitation energy is $E_j^p - E_0$. Because ϕ_{p_j} is selected to interact strongly within the inner radius r_0, the residual orthogonalized set of unoccupied orbitals ϕ_q remain approximately complete in $r_0 \le r \le r_1$ for functions orthogonal to occupied orbitals. The close-coupling equations imply that a closed-channel orbital ϕ_{q_κ} coupled to ϕ_{p_j} is approximately of the form of a slowly decaying function times the open-channel orbital ϕ_κ. Hence the average energy of the two-electron virtual excitations $j, \kappa \to p_j, q$ can be approximated by $E_j^p - E_0$. This argument justifies a closure approximation, and the first line of Eq. (8.16) becomes

$$(\kappa|\hat{v}_c|\kappa) \simeq \sum_j n_j \sum_q (1-n_q)(\kappa j|\bar{u}|q p_j)(q p_j|\bar{c}|\kappa j)$$
$$\simeq -\sum_j n_j \left(\kappa|(j|u|p_j)(E_j^p - E_0)^{-1}(p_j|u|j)|\kappa\right). \qquad (8.17)$$

Table 8.1. *Partial wave phase shifts for He*

	SE		SEP	
k (au)	Present	Ref. [383]	Present	Ref. [383]
$\ell = 0$				
0.3	2.7049	2.7037	2.7424	2.7459
0.5	2.4357	2.4325	2.4812	2.4836
0.7	2.1943	2.1963	2.2433	2.2518
0.9	1.9836	1.9808	2.0339	2.0426
$\ell = 1$				
0.3	0.0140	0.0108	0.0340	0.0297
0.5	0.0447	0.0426	0.0866	0.0847
0.7	0.0980	0.0947	0.1629	0.1567
0.9	0.1570	0.1552	0.2366	0.2311

This is the diagonal matrix element of a multipole polarization potential. For multipole index $\lambda > 0$, transition matrix elements $(p_j|u|j)$ vary as r^λ for small r and as $1/r^{\lambda+1}$ for $r \gg r_0$, so that the multipole polarization potential varies as $r^{2\lambda}$ and $1/r^{2\lambda+2}$, respectively, in these limits. This approximate polarization potential vanishes as $r \to 0$, eliminating the need for an *ad hoc* cutoff for small r. When $\lambda = 1$, the spherically averaged static dipole polarizability is $\alpha_d = \frac{2}{3}(j|\mathbf{x}|p_j) \cdot (p_j|\mathbf{x}|j)(E_j^p - E_0)^{-1}$, a sum over $2\lambda + 1 = 3$ dipole pseudostates. The isotropic polarization potential implied by Eq. (8.17) takes the well-known form $-\alpha_d/2r^4$ for large r. Due to the orthogonality conditions, this approximation is valid only for the long-range part of bound–free correlation, outside the target charge distribution. The fact that the derived multipole polarization potential vanishes for $r \to 0$ may justify simply adding it to a modeled short-range correlation potential, or fitting at an estimated r_0 to such an effective potential [133].

Table (8.1) shows results of test calculations of e–He partial wave phase shifts, compared with earlier variational calculations [383]. The polarization pseudostate was approximated here for He by variational scaling of the well-known hydrogen pseudostate [76]. The present method is no more difficult to implement for polarization response (SEP) than it is for static exchange (SE).

9

Electron-impact rovibrational excitation of molecules

The principal references for this chapter are:

[49] Burke, P.G. and West, J.B., eds. (1987). *Electron–Molecule Scattering and Photoionization* (Plenum, New York).

[82] Domcke, W. (1991). Theory of resonance and threshold effects in electron–molecule collisions: the projection-operator approach, *Phys. Rep.* **208**, 97–188.

[178] Huo, W.M. and Gianturco, F.A., eds. (1995). *Computational Methods for Electron–Molecule Collisions* (Plenum, New York).

[215] Lane, N.F. (1980). The theory of electron–molecule collisions, *Rev. Mod. Phys.* **52**, 29–119.

[284] Nesbet, R.K. (1996). Nonadiabatic phase-matrix method for vibrational excitation and dissociative attachment in electron–molecule scattering, *Phys. Rev. A* **54**, 2899–2905.

[363] Schneider, B.I., LeDourneuf, M. and Burke, P.G. (1979). Theory of vibrational excitation and dissociative attachment: an R-matrix approach, *J. Phys. B* **12**, L365–L369.

Electron–molecule scattering data, observed experimentally or computed with methodology available as of 1980, was reviewed in detail by Lane [215]. If there were no nuclear motion, electron–molecule scattering would differ from electron–atom scattering only because of the loss of spherical symmetry and because of the presence of multiple Coulomb potentials due to the atomic nuclei. This is already a formidable challenge to theory, exemplified by the qualitative increase in computational difficulty and complexity between atomic theory and molecular theory for electronic bound states. While bound-state molecular computational methods have been extended to fixed-nuclei electron scattering [49, 178], an effective and computationally practicable treatment of rovibrational (rotational and vibrational) excitation requires a significant and historically challenging extension of bound-state theory. The small ratio between electron and nuclear masses is exploited in bound-state theory through the Born–Oppenheimer separation [31], leading to the qualitative physical principle that in the lowest order of a perturbation expansion, transfer of kinetic energy between electronic and nuclear motion can be neglected. In the expansion parameter $(m/M)^{\frac{1}{4}}$, where M is a typical nuclear mass, electronic energy is of zeroeth order, vibrational energy of second order, and rotational energy of fourth order. Far from neglecting these Born–Oppenheimer corrections, a valid theory of electron-impact rovibrational excitation must compute the scattering effects of such terms with quantitative accuracy, because the phenomenon of interest is precisely such an energy transfer between electronic and nuclear motion.

Rotational level spacing is in fact so small for most molecules that it can be resolved only by precise spectroscopic techniques. Typical electron scattering data does not have sufficient energy resolution, and most published data is for rotationally averaged scattering cross sections. Special methods based on an adiabatic theory of slowly moving nuclei are valid under these circumstances, not dependent on variational theory and methods. However, near excitation or scattering thresholds, rotational analysis may play an essential role in interpreting complex scattering phenomena due to dynamical long-range potentials, such as the dipole potential for heteropolar diatomic molecules, quenched by molecular rotation. For these reasons, the present discussion of computational methodology will concentrate on threshold effects characteristic of low-energy electron scattering and on vibrational excitation. The phenomenon of dissociative attachment, when an incident electron is captured by a target molecule, which then dissociates to produce an ionic fragment, is a particular example of such energy transfer, and will be considered here as an application of rovibrational threshold theory.

An electron scattering resonance, in abstract but general mathematical terms, is characterized by a singularity of the scattering matrix when analytically continued to a complex energy value $\epsilon_{res} - \frac{i}{2}\gamma$ for relatively small width parameter $\gamma > 0$ [270]. In simple effective-potential models, a so-called *shape* resonance occurs when an electronic energy level in a potential well is degenerate with an energy continuum outside a potential barrier that confines the well. An electron initially bound in the state corresponding to this energy level leaks out through the potential barrier with a time constant $\hbar/\gamma$. Such poles of the analytically continued scattering matrix can also cause significant energy-dependent Wigner cusp or rounded step structures [427, 32] in scattering cross sections just below or just above the energy threshold at which a continuum becomes energetically accessible. This behavior is characteristic of long-range potentials and does not require a well-defined potential barrier [270]. A typical phenomenon is a *virtual state*, which occurs in simple potential models at the threshold for a continuum with orbital angular momentum $\ell = 0$, hence no centrifugal barrier.

Resonance and threshold structures in fixed-nuclei electron-scattering cross sections are replicated, with characteristic energy shifts, in rovibrational excitation cross sections. One of the most striking examples is provided by the prominent multiple peaks observed in both the electronically elastic and vibrational excitation cross sections for electron scattering by the N_2 molecule [368, 143, 95, 215]. These peaks are associated with a fixed-nuclei resonance at approximately 2 eV, but their width arises from electron–vibrational coupling and their spacing does not correspond precisely to the vibrational level structure of either the neutral target molecule or of a vibrating transient negative ion. A quantitative theory of these excitation structures has been developed through several advances in both formalism and computational technique, ultimately based on variational theory of the interacting electron–vibrational system. Another striking example is the observation of excitation peaks associated with each successive vibrational excitation threshold in electron scattering by dipolar molecules such as HF, HCl, and HBr [346]. Because of the electric dipole moment of such target molecules, if low-energy scattering theory is to be relevant it must provide a detailed analysis of rotational excitation and of rotational screening of the long-range dipole potential.

9.1 The local complex-potential (LCP) model

Considerable understanding of molecular resonance and threshold phenomena has come from the conceptually simple model of an electronic resonance state that determines a complex-valued effective internuclear potential function, characterized by a decay width whenever the real part of the resonance energy lies above the fixed-nuclei potential of the target molecule. The vibrational levels of this complex potential are a first approximation to resonance energies such as those observed in N_2. As shown in the "boomerang" model of Herzenberg [167], the implied lifetime of these model vibrational states is so short that they do not survive a single oscillation of a classical wave packet. Hence a dynamical model arises in which excitation from the ground-state vibrational state of the neutral molecule to a wave-packet state of the transient negative ion is followed by a single vibration out and back, traversing a region of relatively long lifetime. Because the decay width increases rapidly for small internuclear distances, this wave packet decays into vibrational states of the neutral molecule as it makes this return trip, and disappears before completing a full oscillation. With appropriate parametrization, this model gives a convincing explanation of the observed vibrational resonance peak shapes and separations in $e{-}N_2$ scattering [24, 91].

The complex-potential model originated as a qualitative explanation of dissociative attachment [18]. A fixed-nuclei electronic resonance is described by a complex potential whose width goes to zero at the crossing point with the molecular ground-state potential, and remains zero if the resonance potential remains below the target potential for large internuclear separation. In this case, an outgoing wave packet becomes a true bound state of the negative ion after crossing outwards, and dissociates without returning. Excitation and decay processes inside the crossing point are the same as in the boomerang model. An incident electron, captured into excited ionic vibrational levels, creates a vibronic wave packet that moves outwards while decaying into vibrational excitations, but eventually simply crosses outwards as a molecular negative ion that dissociates into stable fragments. Applied to the dissociative attachment of H_2 [16, 19, 415], this parametrized model accurately reproduces available experimental data.

A very significant practical problem with the LCP model is the lack of a unique and well-defined theory of the required resonance state when its decay width becomes large. In the case of H_2^-, serious discrepancies between various parametrized or computed values of the energy and width of the $^2\Sigma_u^+$ shape resonance in the LCP model of dissociative attachment indicate that a more fundamental, first-principles theory is needed to give fully convincing results. The difficulty is that as the decay width becomes large, the concept of a well-defined LCP model becomes questionable, since the scattering resonance fades into the background scattering continuum [271]. Analytic theory implies an energy shift which may qualitatively change the character of the assumed resonance potential curve [59, 82].

9.1.1 The projection-operator method

The semiclassical picture of nuclear motion inherent in the LCP model has a quantum-mechanical foundation that is most directly developed in the projection-operator formalism of Feshbach [115, 116, 270]. Domcke [82] reviews the application of this formalism to resonant effects in electron–molecule scattering. Neglecting the kinetic energy of nuclear motion, $\hat{T}_n$, an electronic resonance at nuclear coordinate(s) q is characterized by a pole of the analytically continued scattering matrix $S(q, \epsilon)$ at complex energy $\epsilon = \epsilon_{res}(q) - \frac{1}{2}i\gamma(q)$, where γ is the decay width of the resonance. The set of values $\epsilon_{res}(q)$ defines the real part of the LCP effective potential curve. The Feshbach formalism in electron–molecule scattering is ordinarily applied to the static-exchange model of fixed-nuclei scattering, coupled to the Hamiltonian H_n for nuclear motion [67, 314, 315, 15]. The electronic wave function is represented by a model state Φ (Slater determinant), whose orbital functions are orthogonalized to a normalized, localized orbital function $\phi_d(q; \mathbf{x})$ that interacts with a continuum orbital wave function $\psi_{\mathbf{k}}$ to produce the scattering resonance. For fixed nuclei, this formalism is exactly the same as resonance theory for atoms, resulting in a complex-valued nonlocal optical potential. It is most directly described by a Green function that is formally orthogonalized to the postulated localized function ϕ_d [270, 82].

In the LCP model of vibrational excitation, parametrized functions $\epsilon_{res}(q)$ and $\gamma(q)$, together with the ground-state potential curve, $V_0(q)$, suffice to determine cross sections averaged over rotational substructure. The full Feshbach formalism requires parametrized transition matrix elements between ϕ_d and orthogonalized background continuum orbitals $\hat{\phi}_k$, as well as $V_0(q)$ and the mean electronic energy $V_d(q)$ of a postulated discrete state ϕ_d. In a one-electron model [82], the Feshbach optical potential is added to the effective electronic Hamiltonian for static exchange. This formalism is readily extended to an orbital-functional theory that includes electronic correlation [290]. The transition element V_{dk}, parametrized as a function of both q and the electronic continuum energy ϵ, models the $(N+1)$-electron matrix element $(\Psi_d|H|\mathcal{A}\Theta_0\psi_k)$, where Ψ_d is a postulated discrete $(N+1)$-electron state, Θ_0 is the target electronic ground state, and ψ_k is a continuum orbital function at energy $\epsilon = \frac{1}{2}k^2$. The full function $\mathcal{A}\Theta_0\psi_k$ must be orthogonal to Ψ_d. Thus the formalism defined by such matrix elements applies to the original case considered by Feshbach, where Ψ_d is a core-excited state such as He:$(1s3s^2)^2S$ at 22.45 eV excitation energy, as well as to the explicitly modeled one-electron attached state.

The vibronic Hamiltonian in the one-electron model is $H = H_0 + V$. The kernels of these operators are

$$h_0 = \phi_d[\hat{T}_n + V_d(q)]\phi_d^* + \int k\,dk\,d\Omega_k\hat{\phi}_k\left[\hat{\phi}_n + V_0(q) + \frac{1}{2}k^2\right]\hat{\phi}_k^*;$$

$$v = \int k\,dk\,d\Omega_k\phi_d V_{dk}(q)\hat{\phi}_k^* + hc.$$

The coupled Schrödinger equations can be projected onto the $\phi_d\cdots\phi_d^*$ subspace by Feshbach partitioning, giving an equation for the coefficient function $\chi_d(q)$ in the component $\phi_d\chi_d(q)$ of the total wave function. The effective Hamiltonian in this equation is $\hat{T}_n + V_d(q) + \hat{V}_{opt}$, which contains an optical potential that is nonlocal in the q-space. This operator is defined by its kernel in the $\phi_d\cdots\phi_d^*$ subspace,

$$v_{opt}(q,q';E) = \int k\,dk\,d\Omega_k V_{dk}(q)\left(E - \frac{1}{2}k^2 - \hat{T}_n - V_0(q) + i\eta\right)^{-1} V_{dk}^*(q').$$

This defines a nonlocal complex energy shift $v_{opt} = \Delta - \frac{1}{2}i\Gamma$ such that

$$\Delta(q,q';E) = \sum_v\!\!\!\!\!\!\int k\,dk\,V_{dk}(q)\chi_v(q)\left[E - \frac{1}{2}k^2 - E_v\right]^{-1}\chi_v^*(q')V_{dk}^*(q'),$$

$$\Gamma(q,q';E) = 2\pi\sum_v\!\!\!\!\!\!\int V_{dk_v}(q)\chi_v(q)\chi_v^*(q')V_{dk_v}^*(q').$$

Here $k_v^2 = 2(E - E_v)$ for the bound or continuum vibrational state indexed by v. Thus the Feshbach formalism implies energy-dependent, nonlocal energy-shift and width functions for a resonance.

Neglecting nonresonant scattering, the resonant contribution to the transition matrix is [82]

$$T(\mathbf{k}_f, v_f; \mathbf{k}_i, v_i) = \left(v_f \middle| V^*_{d\mathbf{k}_f} G(E) V_{d\mathbf{k}_i} \middle| v_i\right),$$

deduced from the Lippmann–Schwinger equation. Here $G(E)$ is the resolvent operator in the $\phi_d \cdots \phi_d^*$ space,

$$G(E) = (E - \hat{T}_n - V_d(q) - \hat{V}_{opt} + i\eta)^{-1},$$

evaluated in the limit $\eta \to 0+$ for outgoing-wave boundary conditions. The transition matrix can be computed as

$$T(\mathbf{k}_f, v_f; \mathbf{k}_i, v_i) = \left(v_f \middle| V^*_{d\mathbf{k}_f} \middle| \chi_d\right),$$

since the projected equation for the coefficient function $\chi_d(q)$ implies $\chi_d(q) = G(E)V_{dk_i}\chi_{v_i}$. This projected equation is

$$\{\hat{T}_n + V_d(q) + \hat{V}_{opt} - H\}\chi_d(q) = -V_{dk_i}\chi_{v_i}(q), \tag{9.1}$$

which is nonlocal in the nuclear coordinates [15].

Although in earlier applications $V_d(q) + \hat{V}_{opt}$ was approximated by an empirical local complex potential (LCP), more recent work has solved this nonlocal equation directly [82], obtaining detailed resonance excitation and near-threshold cross sections in excellent agreement with experimental data. Model calculations, using the nonlocal formalism, indicate that a local approximation can yield accurate results for vibrational excitations [59, 258]. The local model is inadequate for very broad resonances [259] and for near-threshold singularities [84, 86]. The neglect of nonresonant scattering implies significant discrepancies for elastic scattering. Despite the practical success of the projection-operator formalism for resonant scattering, the underlying electronic theory is still restricted to a static-exchange model, augmented by an optical potential derived from empirical functions $V_d(q)$ and $\Gamma_d(q)$ or $V_{dk}(q)$.

9.2 Adiabatic approximations

The small mass ratio m/M suggests various levels of approximation based fundamentally on the inherently small kinetic energy of nuclear motion $\hat{T}_n$. The physical concept that justifies the LCP model is that the electronic wave function adjusts essentially instantaneously to displacements of the nuclei, because electronic velocities are much greater than nuclear velocities in a time-dependent semiclassical model. Thus nuclear motion is described in an adiabatic picture, assuming instantaneous electronic relaxation. A more fundamental rationalization can be based on a generalized independent-particle model, including nuclei as well as electrons. This implies a formally exact theory if electronic correlation and electron–nuclei interactions are included. However, the small mass ratio m/M justifies simplification of the self-consistent mean field that acts on each nucleus even though the nuclear motion is described by quantum mechanics and $\hat{T}_n$ is treated as a Schrödinger operator that acts on rovibrational wave functions. Whether or not this mean field can be reduced to a static electronic potential computed for fixed nuclei depends on the limit of linear response theory for the electrons as $m/M \rightarrow 0$. Thus a complete quantitative theory must use the full variational theory of interacting electrons, but can be expected to justify significant simplifications in the treatment of nuclear motion.

Because molecular rotational kinetic energy and rotational level spacings are small in all cases, a rotationally adiabatic model has been widely and successfully used for electron–molecule scattering [215]. The essence of this *adiabatic nuclei* (ADN) approximation [313, 63, 64, 401] is that a scattering amplitude or matrix computed for fixed nuclei is treated as an operator in the nuclear coordinates. Then the rotational state-to-state scattering amplitude or matrix is estimated by matrix elements of this operator in the basis of rotational states. The basic assumption is that of the Born–Oppenheimer separation: the commutator between the fixed-nuclei scattering operator and the operator $\hat{T}_n$ is neglected. This approximation becomes difficult to justify only when the precise energy of individual rotational levels is important, near thresholds or in the presence of long-range potentials. Because of the well-defined internuclear geometry in a body-fixed reference frame, fixed-nuclei scattering matrices are most directly computed in this frame. In contrast, rotational scattering structure is observed in a fixed laboratory frame. Unless an explicit rotational frame transformation is carried out, strong rotational coupling occurs as a computational artifact [215]. For this reason, applications of the rotational ADN model require such a frame transformation [62].

The ADN approximation has been much less successful for vibrational excitation, as might be expected from the much larger vibrational level spacings and the strong variation of fixed-nuclei resonance parameters with nuclear displacements. This motivated the proposal [61] of a *hybrid* close-coupling model. As applied to $e-N_2$ vibrational excitation, this model combines fixed-nuclei electronic close-coupling calculations for nonresonant body-frame molecular symmetry states with extended vibronic close-coupling calculations in the $^2\Pi_g$ resonant symmetry of N_2^-. Although multipeaked vibrational excitation and vibrationally elastic cross sections are obtained in qualitative agreement with experiment, the peak shapes and spacings are not in good agreement. This can be attributed to truncation of the electronic partial-wave expansion and of the limited number of vibrational states included in the close-coupling basis. A deeper problem is that any complete set of vibrational states, such as the eigenfunctions of a parametrized Morse potential, must include the vibrational continuum. No practical way has been found to do this in the close-coupling formalism.

9.2.1 The energy-modified adiabatic approximation (EMA)

In many scattering processes, energy levels of the target system are split by a perturbation that is weak relative to the interaction responsible for the scattering. The adiabatic approximation neglects the effect on threshold scattering structures and on resonances of the energy-level splitting and energy shifts of the perturbed target states. Especially for threshold structures, this can lead to qualitatively incorrect results. For example, transition matrix elements computed below a rovibrational excitation threshold do not vanish if only the fixed-nuclei target energy is taken into account. This qualitatively incorrect behavior of adiabatic cross sections near rotational excitation thresholds can be compensated simply by modifying electron momenta in the adiabatic cross-section formula [64]. With these corrections, the ADN theory appears to be adequate for rotational excitation [144].

Much larger anomalous effects occur in vibrational excitation because the energy shifts are larger. The energy-modified adiabatic approximation (EMA) [267] was introduced in order to provide a systematic treatment of such effects, while retaining the computational efficiency of the adiabatic approximation. The usual adiabatic approximation is modified by allowing for the dependence of unperturbed scattering matrices on the kinetic energy of the perturbed target state, using formulas that are qualitatively correct for threshold and resonance structures. For molecules, this means that for target geometry determined by a generalized coordinate q the energy $\epsilon(q) = E - V(q)$ of a continuum electron must be replaced in principle by $\epsilon_{op}(q) = E - H_n(q)$, an operator in the nuclear coordinates. Fixed-nuclei calculations produce scattering matrices that are functions of numerical parameters q, ϵ. In the EMA, the integrals that project these matrices onto rovibrational states are evaluated by approximations that replace the parameter ϵ by the operator ϵ_{op}. The simplest such approximation, for diagonal matrix elements in a rovibronic state indexed by μ, is to replace ϵ by $\epsilon_\mu = E - E_\mu$. If nondiagonal matrix elements between states indexed by μ and ν are evaluated for the geometric mean energy $\epsilon_{\mu\nu} = [(E - E_\mu)(E - E_\nu)]^{\frac{1}{2}}$, this implies correct threshold behavior for general short-range potentials. When applied to the energy-denominator characteristic of a fixed-nuclei molecular resonance, this state-dependent modification of ϵ to ϵ_{op} was shown to give qualitatively correct results for the multipeaked vibrational excitation structures observed in e$-N_2$ scattering [267]. Because the EMA formalism replaces fixed-nuclei scattering matrices by operators that are represented by rovibronic scattering matrices, a single fixed-nuclei or threshold structure becomes a set of overlapping scattering structures, displaced with possibly irregular energy shifts by the discrete vibronic energy level structure of the target molecule [267]. This repetition and displacement of underlying structures is characteristic of observed electron scattering cross sections for molecular targets. A more recent extension of the EMA formalism to the context of variational R-matrix theory is discussed below.

9.3 Vibronic *R*-matrix theory

Recognizing that exact quantum electron–molecule scattering theory for interacting nuclei and electrons is and will remain computationally intractable, except for the simplest diatomic molecules, Schneider [361] initiated reconsideration of the Born–Oppenheimer approximation as a logical foundation for the adiabatic nuclei (ADN) formalism. For a long-lived (narrow) resonance, neglect of vibrational derivatives of the fixed-nuclei electronic wave function should be no less valid for electron scattering at low energies than it is for molecular bound states. This argument is the basis of an electronic R-matrix methodology in which eigenstates, defined by eigenvalues of the Bloch-modified electronic Hamiltonian from which the R-matrix is constructed, are used to define effective molecular potential functions, parametrized by nuclear coordinates [363]. The small mass ratio m/M justifies neglecting derivatives of the electronic eigenfunctions of this fixed-nuclei Bloch-modified Hamiltonian with respect to the nuclear coordinates.

This work introduced the concept of a vibronic R-matrix, defined on a hypersurface in the joint coordinate space of electrons and internuclear coordinates. In considering the vibronic problem, it is assumed that a matrix representation of the Schrödinger equation for $N+1$ electrons has been partitioned to produce an equivalent set of multichannel one-electron equations coupled by a matrix array of nonlocal optical potential operators [270]. In the body-fixed reference frame, partial wave functions in the separate channels have the form $\Theta_p(q;\mathbf{x}^N)Y_L(\theta,\phi)\chi_v(q)$, multiplied by a radial channel orbital function $\psi(q;r)$ and antisymmetrized in the electronic coordinates. Here Θ is a fixed-nuclei N-electron target state or pseudostate and Y_L is a spherical harmonic function. Both Θ and ψ are parametric functions of the internuclear coordinate q. It is assumed that the target states Θ for each value of q diagonalize the N-electron Hamiltonian matrix and are orthonormal.

An electronic R-matrix radius a is chosen such that exchange can be neglected for $r > a$. An upper limit q_d for the internuclear coordinate q is chosen so that a dissociating electronic state Φ_d is bound for $q \geq q_d$. This defines a vibronic hypercylinder [284] with two distinct surface regions: an electronic wall with $r = a$ for $0 \leq q \leq q_d$ and a dissociation cap defined by the enclosed volume of the electronic sphere for $q = q_d$. For nondissociating molecules, q_d should be large enough to enclose the highest vibrational state to be considered. The R-matrix is defined by matrix elements of the variational operator $\mathcal{R}$ in a complete basis of surface functions [284], such as the spherical harmonics on the wall surface.

A matrix of operators $\mathcal{R}_{pp'}$ is defined by projection of $\mathcal{R}$ into the multichannel representation indexed by N-electron target states Θ_p. This defines the vibrational excitation submatrix of the R-matrix as

$$(pLv|R|p'L'v') = \int dq \int dq' \int d\Omega \int d\Omega' \chi_v^*(q) Y_L^*(\theta\phi) \mathcal{R}_{pp'}(q\theta\phi; q'\theta'\phi') Y_{L'}(\theta'\phi') \chi_{v'}(q').$$

This matrix can be computed from the general variational formula derived in Chapter 8, using a complete set of vibronic basis functions

$$\Psi_\alpha(\mathbf{x}^{N+1}, q) = \mathcal{A}\Theta_p(q; \mathbf{x}^N) Y_L(\theta, \phi) \psi_\alpha(q; r) \chi_\alpha(q).$$

It is assumed that target states Θ_p are indexed for each value of q such that a smooth diabatic energy function $E_p(q)$ is defined. This requires careful analysis of avoided crossings. The functions χ_α should be a complete set of vibrational functions for the target potential $V_p = E_p$, including functions that represent the vibrational continuum. All vibrational basis functions are truncated at $q = q_d$, without restricting their boundary values. The radial functions ψ_α should be complete for $r \leq a$.

Free boundary conditions are not allowed in the formulation of the theory by Schneider *et al.* [363], based on the nonvariational theory of Wigner and Eisenbud [428]. Specific boundary conditions are imposed using a Bloch operator. This determines boundary conditions correctly at energy poles of the R-matrix determinant, but requires a Buttle correction [54] for energy values between such poles [363]. This becomes problematic for the internuclear coordinate, because the physical model of the dissociating state is a complex potential function for $q \leq q_d$, so that fixed boundary conditions imply complex energy eigenvalues. Nevertheless, in the usual case that R-matrix poles are associated with homogeneous Neumann boundary conditions on the R-matrix boundary, the Wigner–Eisenbud theory and variational R-matrix theory derive the same equations for the vibronic R-matrix.

The projection integrals on the electronic wall are

$$(pLv|\alpha) = (\mathcal{A}\Theta_p \chi_v|\Psi_\alpha)_{r=a} = \int_0^{q_d} \chi_v^*(q) \psi_\alpha(q; a) \chi_\alpha(q)\, dq.$$

In agreement with [363], the R-matrix for vibrational excitation is

$$(pLv|R|p'L'v') = \frac{1}{2} \sum_{\alpha,\beta} (pLv|\alpha)(H_B - E)^{-1}_{\alpha\beta}(\beta|p'L'v'),$$

where H_B is the Bloch-modified vibronic Hamiltonian. This requires vibrational kinetic energy matrix elements to be evaluated as the Hermitian form $\frac{1}{2\mu}\int dq \frac{d\chi_v^*}{dq}\frac{d\chi_{v'}}{dq}$.

The derivation up to this point involves no approximations if the vibronic basis set is complete in the closed hypervolume, including its surface. If a dissociation channel exists, it can be approximated by projection onto a single diabatic state $\Phi_d(q; \mathbf{x}^{N+1})$, assumed to be well defined as a discrete state on the cap surface $q = q_d$. The projection integrals on this surface are

$$(d|\alpha) = (\Phi_d|\Psi_\alpha)_{q=q_d}.$$

The R-matrix connecting wall and cap surfaces, obtained by projecting the $\mathcal{R}$ operator onto both surfaces, is

$$(d|R|pLv) = \frac{1}{2}\sum_{\alpha,\beta}(d|\alpha)(H_B - E)^{-1}_{\alpha\beta}(\beta|pLv),$$

which determines excitation–dissociation transitions. The R-matrix in the cap surface is

$$(d|R|d) = \frac{1}{2}\sum_{\alpha,\beta}(d|\alpha)(H_B - E)^{-1}_{\alpha\beta}(\beta|d).$$

This is the reciprocal of the logarithmic derivative of the wave function $\chi_d(q)$ in the dissociation channel, for $q = q_d$. At given total energy E these R-matrix elements are matched to external scattering wave functions by linear equations that determine the full scattering matrix for all direct and inverse processes involving nuclear motion and vibrational excitation. Because the vibronic R-matrix is Hermitian by construction (real and symmetric by appropriate choice of basis functions), the vibronic S-matrix is unitary.

Schneider *et al.* [363] use Born–Oppenheimer vibronic basis functions as indicated above, and neglect Born–Oppenheimer corrections determined by the internuclear momentum operator acting on the electronic wave function. Radial basis functions $\psi_k(q; r)$ correspond to R-matrix pole states, whose energy values $E_k(q)$ define an indexed vibrational potential. Vibrational basis functions $\chi_{k\mu}(q)$ are computed as eigenfunctions of the corresponding Hamiltonian. Since resonance states are not treated by projection, the method depends on effective completeness of the double expansion in electronic and vibrational eigenfunctions. The first application of this method, to the multipeaked vibrational excitation structure observed in $e{-}N_2$ scattering, was remarkably successful [364], in much closer agreement with experiment than were comparable calculations using the hybrid ADN close-coupling formalism [61]. Electronic wave functions with fixed boundary conditions at $r = a$ were used for the four lowest R-matrix pole states, and vibrational wave functions were computed without considering the vibrational continuum. Boundary values on the cap surface are not relevant since dissociation is not involved. A Buttle correction for the electronic basis was computed using adiabatic theory.

These calculations were later extended up to 30 eV [138] scattering energy, including differential cross sections for elastic scattering and vibrational excitation. The original method was modified to use a fixed electronic basis set for all internuclear distances, in order to mitigate problems arising from avoided crossings of R-matrix pole state potential curves. Vibrational wave functions were represented in a basis of shifted Legendre polynomials. These procedures were used for calculations of integral and differential cross sections in e–HF scattering [250] and for similar calculations on HCl [251], both examples of threshold excitation peaks due to the molecular dipole moment. Because results computed with this method, based on a fixed boundary condition at $q = q_d$, depend strongly on the choice of q_d, a modified theory has been proposed in which the evidently successful R-matrix theory of vibrational excitation is combined with resonance-state theory for nuclear motion [103]. The approximation of neglecting derivatives of electronic wave functions with respect to internuclear coordinates appears to be satisfactory in all of these applications [362].

9.3.1 Phase-matrix theory

The general success of projection-operator methods indicates that quantitative calculations can be based on the strategy of separating singularities of scattering matrices from a smoothly varying background [82]. The R-matrix is in general a real symmetric matrix with isolated real energy poles, analogous to the K-matrix in scattering theory. For the K-matrix, these poles have no special physical significance, simply indicating that the sum of eigenphases of the corresponding unitary S-matrix passes through an odd multiple of $\pi/2$ radians. Similarly, the choice of pole states of the R-matrix to define "vibrational" potential functions in the method of Schneider *et al.* [363] is an arbitrary construction whose principal effect is to produce linearly independent vibronic basis states that can be extended to a complete set inside the R-matrix hypersurface.

In order to establish a better-motivated connection to resonance theory, the fixed-nuclei R-matrix can be converted to a phase matrix Φ, defined such that $\tan \Phi = k(q)R$, or to the corresponding unitary matrix

$$U = (I + ik(q)R)/(I - ik(q)R),$$

the analog of the scattering S-matrix, whose complex energy poles define scattering resonances and bound states. The factor $k(q) = [2(E - V_0(q))]^{\frac{1}{2}}$ makes Φ dimensionless. A resonance is characterized for real energies by a point of most rapid increase of the eigenphase sum [389, 270], which is the trace of the matrix $\tan^{-1} K$.

Evaluated at specified $r = a$, the phase matrix Φ of the R-matrix has analogous properties. For real energies, Φ has a monotonically increasing trace, which can be made continuous by suitable choice of the branch of each multivalued eigenphase function, adding or subtracting integral multiples of π at each energy value. A point of maximum slope defines a "precursor resonance" [284] independently of any specific physical model, which corresponds to a pole of the analytically continued S-matrix as the R-matrix boundary is increased. The proposed methodology uses time-delay analysis [389] to separate a given fixed-nuclei phase matrix into a rapidly varying resonant part and a slowly varying background part. The rovibronic phase matrix is constructed by applying resonance theory to the resonant phase matrix only, treating the nonresonant background part by energy-modified adiabatic theory. By separating out rapid variations of Φ this methodology reduces the completeness requirement from the double expansion inherent in the method of Schneider *et al.* [363] to a single expansion for a well-defined precursor resonance state. Vibrational completeness is obtained by using a special basis of "spline-delta" functions that make no distinction between bound and continuum vibrational wave functions [295].

9.3.2 Separation of the phase matrix

With current computational methods, accurate fixed-nuclei R-matrices R^{FN} can be obtained that interpolate smoothly in a vibrational coordinate q and in the electronic continuum energy ϵ. The fixed-nuclei phase matrix Φ^{FN} is defined such that

$$\tan \Phi^{FN}(q;\epsilon) = k(q)R^{FN}(q;\epsilon),$$

where eigenphases are adjusted by multiples of π to make matrix elements continuous in both ϵ and q. For the vibronic phase matrix,

$$\tan \Phi_{vv'} = k_v^{\frac{1}{2}} R k_{v'}^{\frac{1}{2}},$$

where $k_v = [2(E - E_v)]^{\frac{1}{2}}$. A precursor resonance corresponds to rapid variation of the trace of the phase matrix. Single-pole parameters for a resonance can be determined from the energy derivative of the phase matrix [389]. Analysis of a single-channel resonance [270] shows that an S-matrix pole at complex energy $\epsilon_{res} - \frac{1}{2}i\gamma$ implies that the energy derivative of the phase shift $\eta(\epsilon)$ has a maximum value at ϵ_{res}. Assuming constant background phase variation, $\epsilon_{res}(q) = E_{res}(q) - V_0(q)$ is defined by a local maximum of $\frac{d}{d\epsilon}Tr\,\Phi$. The maximum eigenvalue of the matrix $\frac{d}{d\epsilon}\Phi$ at ϵ_{res} is $2/\gamma$ [389]. The eigenvector y defines a resonance eigenchannel. This analysis determines the parameters in the Breit–Wigner formula for an isolated multichannel resonance [270], consistent with a single pole of the analytically continued U-matrix as defined above. This implies an analytic formula for the resonant phase matrix Φ_1, which will be considered in more detail below. Given the phase matrix Φ, this construction of Φ_1 defines a background matrix Φ_0 by subtraction, such that $\Phi = \Phi_0 + \Phi_1$. This procedure can be repeated for several neighboring precursor resonances if necessary. Since Φ is unchanged, no information is lost.

9.3.3 Phase-matrix formalism: EMAP

After separation into resonant and background parts, the nonresonant fixed-nuclei phase matrix Φ_0 is converted to a vibronic or rovibronic phase matrix by the energy-modified adiabatic phase-matrix method (EMAP) [409]. This is simply an adaptation of the EMA formalism to the phase matrix Φ_0. The implied vibronic background phase matrix is

$$(pLv|\Phi_0|p'L'v') = \int \chi_v^*(q)\big(pL\big|\Phi_0^{FN}(q;\epsilon_{vv'})\big|p'L'\big)\chi_{v'}(q)\,dq,$$

where $\epsilon_{vv'} = [(E - E_v)(E - E_{v'})]^{\frac{1}{2}}$. The geometric mean is appropriate to the dimensionless vibronic phase matrix for general short-range potentials [236], as in the earlier EMA theory [267].

The EMAP method has been used in *ab initio* calculations of near-threshold rotational and vibrational excitation in electron scattering by polar molecules [409, 410]. Computed differential cross sections are in quantitative agreement with available experimental data for e−HF scattering. Because of the computational simplifications resulting from combining R-matrix theory with the adiabatic approximation, this methodology was able to obtain results equivalent to converged close-coupling calculations including both vibrational and rotational degrees of freedom. Specific treatment of rotational structure is essential for such molecules because of dynamical rotational screening of the long-range dipole potential. These calculations provide a detailed analysis of the striking threshold peak structures observed for such dipolar molecules. For e−H_2 vibrational excitation, in a direct comparison with the FONDA (first-order nondegenerate adiabatic) approximation [252, 2, 254], and with benchmark vibrational close-coupling results [411], the EMAP method was found to be computationally efficient and reasonably accurate at energies somewhat above threshold and away from a scattering resonance [234].

The EMAP method has been used to compute elastic scattering and symmetric-stretch vibrational excitation cross sections for electron scattering by CO_2 [235]. This is one of the first *ab initio* calculations of vibrational excitation for a polyatomic molecule. The results are in good agreement with experiment, which shows unusually large low-energy cross sections. The theory identifies a near-threshold singularity in the fixed-nuclei scattering matrix, changing from a virtual state to a bound state as the vibrational coordinate varies [249, 235].

9.3.4 Nonadiabatic theory: NADP

For low collision energies, especially when electronic resonances occur, and for processes such as dissociative attachment, adiabatic theory is not adequate for vibrational excitation and energy transfer. The strategy of separating the fixed-nuclei phase matrix into resonant and nonresonant parts makes it possible to apply resonance analysis, analogous to the very successful projection-operator method [82], to the rapidly varying part of the phase matrix. This nonadiabatic phase-matrix (NADP) formalism is derived here. Since adiabatic theory is generally adequate for rotational coupling and excitation, the discussion here is limited to the model considered above, specifically for a diatomic molecule in the body-frame, with only one internuclear coordinate q.

As described above, time-delay analysis [389] of the energy derivative of the phase matrix Φ determines parametric functions that characterize the Breit–Wigner formula for the fixed-nuclei resonant R-matrix $R_1^{FN}(q;\epsilon)$. The resonance energy $\epsilon_{res}(q)$, the decay width $\gamma(q)$, and the channel-projection vector $y(q)$ define R_1^{FN} and its associated phase matrix Φ_1^{FN}, such that $\tan \Phi_1^{FN} = k(q)R_1^{FN}$, where

$$R_1^{FN}(q;\epsilon) = \frac{1}{2} y(q)\gamma^{\frac{1}{2}}(q)[\epsilon_{res}(q) - \epsilon(q)]^{-1}\gamma^{\frac{1}{2}}(q)y^{\dagger}(q). \tag{9.2}$$

Using the basic rationale of EMA theory [267], the parametric function $\epsilon(q)$ becomes $\epsilon_{op}(q) = E - H_n$ when the kinetic energy of nuclear motion cannot be neglected. However, the operator $(\epsilon_{res}(q) - \epsilon_{op}(q))$ has a well-defined c-number value in vibrational eigenstates determined by the eigenvalue equation

$$(\hat{T}_n + E_{res}(q))\chi_s(q) = \chi_s(q)E_s. \tag{9.3}$$

Defining $\epsilon_s(q) = E_s - V_0(q)$, this implies

$$(\epsilon_{res}(q) - \epsilon_{op}(q))\chi_s(q) = \chi_s(q)(\epsilon_s(q) - \epsilon(q)).$$

Here $\epsilon_s(q) - \epsilon(q) = E_s - E$, independent of q. Thus the energy denominator in Eq. (9.2) can be replaced by $E_s - E$ in the vibrational eigenstate χ_s. In the NADP method, Eq. (9.3) is solved in a basis of spline delta-functions [295], which determines bound and continuum vibrational eigenfunctions to graphical accuracy (a cubic spline fit) in the coordinate range $0 \leq q \leq q_d$. Substituting this c-number energy denominator into Eq. (9.2), and evaluating matrix elements of the resulting operator in the vibrational coordinate q, the NADP resonant vibronic R-matrix for vibrational excitation is

$$(pLv|R_1|p'L'v') = \frac{1}{2}\sum_s \left(v\middle|y_{pL}\gamma^{\frac{1}{2}}\middle|s\right)(E_s - E)^{-1}\left(s\middle|\gamma^{\frac{1}{2}}y^{\dagger}_{p'L'}\middle|v'\right). \quad (9.4)$$

For comparison with projection-operator theory, this corresponds to a Born–Oppenheimer precursor resonance state

$$\Psi_s(\mathbf{x}^{N+1}, q) = \Phi_d(q; \mathbf{x}^{N+1})\chi_s(q),$$

where Φ_d is a postulated discrete state, defined for $q \leq q_d$, that interacts with the background electronic continuum. For this precursor state, the projection integrals on the electronic wall of the vibronic hypercylinder are $(pLv|s) = (\mathcal{A}\Theta_p Y_L \chi_v|\Psi_s)_{r=a}$, and the corresponding projection integral on the dissociation cap is $(d|s) = (\Phi_d|\Psi_s)_{q=q_d} = \chi_s(q_d)$, a normalized eigenfunction of Eq. (9.3). The vibronic R-matrix for this precursor resonance state is

$$(pLv|R_{res}|p'L'v') = \frac{1}{2}\sum_s (pLv|s)(E_s - E)^{-1}(s|p'L'v'), \quad (9.5)$$

the R-matrix connecting wall and cap surfaces is

$$(d|R_{res}|pLv) = \frac{1}{2}\sum_s \chi_s(q_d)(E_s - E)^{-1}(s|pLv),$$

and the R-matrix on the cap surface is

$$(d|R_{res}|d) = \frac{1}{2}\sum_s \chi_s(q_d)(E_s - E)^{-1}\chi_s^*(q_d).$$

As a consistency check, if a Green function is defined by

$$G(q, q') = \frac{1}{2}\sum_s \chi_s(q)(E_s - E)^{-1}\chi_s^*(q'),$$

all formulas agree with Schneider *et al.* [363].

Comparing Eqs. (9.5) and (9.4), they would be identical if the "magic formula"

$$(pLv|s) \cong \left(v\middle|y_{pL}\gamma^{\frac{1}{2}}\middle|s\right)$$

were valid. The NADP formalism postulates this to be true, implying that the R-matrix connecting wall and cap surfaces is

$$(d|R_1|pLv) = \frac{1}{2}\sum_s \chi_s(q_d)(E_s - E)^{-1}\left(s|\gamma^{\frac{1}{2}}y^{\dagger}_{pL}|v\right),$$

and the R-matrix on the cap surface is

$$(d|R_1|d) = \frac{1}{2}\sum_s \chi_s(q_d)(E_s - E)^{-1}\chi_s^*(q_d).$$

Because the postulated diabatic state Φ_d is never computed explicitly, the NADP formalism completely avoids the conceptual difficulties associated with this state in the LCP and projection operator methods. Reduction of the vibrational computation to solution of Eq. (9.3) removes the difficult issue of vibrational completeness inherent in vibrational close-coupling theory and in the method of Schneider *et al.*

The NADP method was first tested in calculations of e$-$$N_2$ rovibrational excitation. The efficiency of this formalism was demonstrated by carrying the calculations to effective completeness for combined rotational and vibrational close-coupling. The multipeaked vibrational excitation structure was computed to an accuracy that appears to agree more closely with experiment than does any previous theoretical work [153]. Computed differential vibrational excitation cross sections are in close agreement with experiment. More recently, the NADP method has been used in a series of calculations of e$-$$H_2$ scattering intended to calibrate the method against earlier work that was designed to give definitive results for low-energy vibrational excitation cross sections. The $^2\Pi_u$ shape resonance of H_2^-, which dominates electron scattering for energies below 10 eV, has traditionally been very difficult to characterize, because the indicated decay width becomes very large at internuclear distances near the the ground-state equilibrium q_0. It was found in NADP calculations [236, 233] that the precursor resonance considered in this methodology is in fact very well defined, and the resulting resonance vibrational excitation cross sections are in close agreement with the best available prior calculations.

IV

Field theories

This part is concerned with variational principles underlying field theories. Chapter 10 develops the nonquantized theory of interacting relativistic fields, emphasizing Lorentz and gauge invariant Lagrangian formalism. The theory of a classical nonabelian gauge field is carried to the point of proving gauge invariance and of deriving the local conservation law for field energy and momentum densities.

10
Relativistic Lagrangian theories

The principal references for this chapter are:

[73] Cottingham, W.N. and Greenwood, D.A. (1998). *Introduction to the Standard Model of Particle Physics* (Cambridge University Press, New York).
[93] Edelen, D.G.B. (1969). *Nonlocal Variations and Local Invariance of Fields* (American Elsevier, New York).
[121] Feynman, R.P. (1961). *Quantum Electrodynamics* (Benjamin, New York).
[217] Leader, E. and Predazzi, E. (1996). *An Introduction to Gauge Theories and Modern Particle Physics*, Vols. 1 and 2 (Cambridge University Press, New York).
[336] Ramond, P. (1989). *Field Theory: A Modern Primer,* 2nd edition (Addison-Wesley, Redwood City, California).
[340] Renton, P. (1990). *Electroweak Interactions* (Cambridge University Press, New York).
[352] Ryder, L.H. (1985). *Quantum Field Theory* (Cambridge University Press, New York).
[373] Schwinger, J. (1958). *Quantum Electrodynamics* (Dover, New York).
[391] Swanson, M.S. (1992). *Path Integrals and Quantum Processes* (Academic Press, New York).
[421] Weinberg, S. (1995). *The Quantum Theory of Fields*, Vol. I (Cambridge University Press, New York).
[422] Weinberg, S. (1996). *The Quantum Theory of Fields*, Vol. II (Cambridge University Press, New York).

In quantum electrodynamics (QED), the classical electromagnetic field A_μ of Maxwell and the electronic field ψ of Dirac are given algebraic properties (Bose–Einstein and Fermi–Dirac quantization, respectively), and through their interaction account for almost all physical phenomena that can be observed in ordinary human circumstances. The relativistic theory is derived from Hamilton's principle for an action defined by the space-time integral of a Lorentz invariant Lagrangian density [373]. This same action integral can be used to develop the diagrammatic perturbation theory of Feynman [121]. The cited references describe the formalism and methodology which demonstrate that QED is in remarkable agreement with all empirical data to which it is applicable. Classical and quantized QED will be used here to introduce the basic formalism of field theory, including the variational theory of invariance properties. This theory, especially gauge invariance, is central to recent developments of electroweak theory (EWT) and quantum chromodynamics (QCD).

Because both the Maxwell and the Dirac fields appear to be truly elementary, not simply approximate models of inaccessible underlying structures, QED serves as a model for a more general field theory that might ultimately describe all phenomena. In recent years, some of the mystery has been removed from nuclear and high-energy physics by the discovery (or invention) of quarks and gauge boson fields, as analogs and generalizations of electrons and the electromagnetic field, respectively. Starting from QED as a generic gauge theory, the formalism of electroweak theory (EWT) is developed here, following the implications of nonabelian gauge symmetries in field theory. The same principles apply to the quantum chromodynamics (QCD) of quarks and gluons, which will not be considered in detail.

10.1 Classical relativistic electrodynamics

The term "classical" here implies "nonquantized", treating the Dirac spinor field ψ as a classical field in space-time. The theory is simplified by considering a representation in terms of truly elementary unrenormalized fields, corresponding to an electron with no "bare" mass, and a classical Maxwell field, interacting through an unrenormalized coupling constant. Physically observed electronic mass and charge can be considered to be dynamical results of such a model, when extended by field quantization and augmented by additional terms in the Lagrangian density. The bare fermion field ψ is described by a Pauli 2-spinor, with definite left-handed helicity [25, 336]. Elementary massless fermions with left-handed helicity have positive energy, and those with right-handed helicity have negative energy. In the bare vacuum state of the quantized theory, all right-handed chiral states are fully occupied. Holes in these states are equivalent to right-handed antifermion states of positive energy and reversed charge. Thus the massless QED model includes electrons and positrons from the outset. The effect of renormalization, to be considered below, is to mix bare left-handed electrons of positive energy with bare right-handed electrons of negative energy, to produce the "dressed" electron described by the Dirac 4-component spinor, with the observed mass and charge by construction. In comparison with EWT and QCD, this model of QED neglects only weak interactions, since electrons do not interact with the gluon fields.

The QED Lagrangian density in mixed Gaussian units is

$$\mathcal{L} = -(1/16\pi)F^{\mu\nu}F_{\mu\nu} + i\hbar c\psi^\dagger\gamma^0\gamma^\mu D_\mu\psi,$$

using the customary summation convention for repeated indices. Covariant 4-vectors are defined here by

$$x_\mu = (ct, -\mathbf{r}), \qquad \partial_\mu = (\partial/c\partial t, \boldsymbol{\nabla}),$$
$$A_\mu = (\phi, -\mathbf{A}), \qquad j_\mu = (c\rho, -\mathbf{j}).$$

The signs of spatial components are reversed in the corresponding contravariant 4-vectors, indicated by x^μ, etc. Dirac matrices are represented in a form appropriate to a 2-component fermion theory [38], in which helicity γ^5 is diagonal for zero-mass fermions,

$$\gamma^\mu = \begin{pmatrix} 0 & I \\ I & 0 \end{pmatrix}, \begin{pmatrix} 0 & \boldsymbol{\sigma} \\ -\boldsymbol{\sigma} & 0 \end{pmatrix}; \quad \gamma^5 = \begin{pmatrix} -I & 0 \\ 0 & I \end{pmatrix}. \tag{10.1}$$

The covariant electromagnetic field is defined in terms of the electromagnetic 4-potential A_μ by the antisymmetric form

$$F_{\mu\nu} = \partial_\mu A_\nu - \partial_\nu A_\mu.$$

The Dirac and Maxwell fields are coupled through the covariant derivative

$$D_\mu = \partial_\mu + i(-e/\hbar c)A_\mu,$$

where $-e$ is the electronic charge.

QED theory is based on two distinct postulates. The first is the dynamical postulate that the integral of the Lagrangian density over a specified space-time region is stationary with respect to variations of the independent fields A_μ and ψ, subject to fixed boundary values. The second postulate attributes algebraic commutation or anticommutation properties, respectively, to these two elementary fields. In the classical model considered here, the dynamical postulate is retained, but the algebraic postulate and its implications will not be developed in detail.

The dynamical postulate implies the covariant field equations,

$$\partial^\mu F_{\mu\nu} = (4\pi/c)j_\nu = (4\pi/c)(-ec\psi^\dagger\gamma^0\gamma_\nu\psi),$$
$$i\hbar c\gamma^\mu D_\mu\psi = 0. \tag{10.2}$$

Expanded in a complete set of spinor functions, the fermion field is $\psi(x) = \sum_p u_p(x)a_p$, where $x = (\mathbf{x}, t)$. In a particular Lorentz frame, Eq. (10.2) takes the form

$$i\hbar\partial_t\psi = \mathcal{H}\psi = -i\hbar\{c\gamma^0\gamma^\mu D_\mu - \partial_t\}\psi. \tag{10.3}$$

For zero rest mass, using Eqs. (10.1), the 2×2 operator that acts on 2-component spinors of negative helicity is

$$\mathcal{H}_L = \boldsymbol{\sigma}\cdot(i\hbar c\boldsymbol{\nabla} - e\mathbf{A}) - eA_0$$
$$= -c\boldsymbol{\sigma}\cdot\left(\mathbf{p} + \frac{e}{c}\mathbf{A}\right) - e\phi.$$

10.1.1 Classical dynamical mass

If there is no explicit external electromagnetic field, the covariant field equations determine a self-interaction energy that can be interpreted as a dynamical electron mass δm. Since this turns out to be infinite, renormalization is necessary in order to have a viable physical theory. Field quantization is required for quantitative QED. The classical field equation for the electromagnetic field can be solved explicitly using the Green function or Feynman propagator $G_P^{\mu\nu}$, whose Fourier transform is $-g^{\mu\nu}/\kappa^2$, where $\kappa = k_p - k_q$ is the 4-momentum transfer. The product of γ_0 and the field-dependent term in the Dirac Hamiltonian, Eq. (10.3), is

$$-e\not{A}(x) = -e\gamma^\mu A_\mu(x)$$
$$= 4\pi e^2\gamma_\mu \int d^4x' G_P^{\mu\nu}(x-x')\sum_{p,q} a_p^\dagger u_p^*(x')\gamma^0\gamma_\nu u_q(x')a_q.$$

In terms of the 4-momentum transfer κ this is

$$-e\not{A}(x) = -4\pi e^2\gamma_\mu \int \frac{d^4\kappa}{(2\pi)^4}\frac{g^{\mu\nu}}{\kappa^2}\sum_{p,q} a_p^\dagger \int d^4x' e^{-i\kappa\cdot(x'-x)} u_p^*(x')\gamma^0\gamma_\nu u_q(x')a_q.$$

This is an exact result of the theory, and must retain its validity for quantized fields. If there is no external field, any scalar field resulting from coupling $u_p^* u_q$ must vanish. This is required in the vacuum state of the quantized theory, and a valid classical model should omit such terms. Because the functions $u_p(x)$ are elementary Pauli spinors, the nonscalar field $\not{A}(x)$ has triplet-odd character, describing a vector boson in quantized theory. Then $\not{A}(x)$ is an odd-parity field, whose matrix elements vanish except for transitions between negative and positive energies, because massless fermions have definite helicity. Chirality-breaking virtual transitions are required in order to produce a nonvanishing mass from the electromagnetic interaction.

In relativistic theory, it is desirable to retain covariant forms in the equations of motion. Instead of Eq. (10.3), Eq. (10.2) can be rewritten as

$$i\hbar c\gamma^\mu\partial_\mu\psi = -e\not{A}\psi = \delta\hat{m}c^2\psi,$$

defining $\delta\hat{m}c^2 = -e\not{A}$. This is an operator or matrix in the Pauli spinor representation, since its nonvanishing matrix elements describe transitions between helicity states. If there were a bare mass m_0, a term $m_0c^2\psi$ would be added to the right-hand side of this equation, defining $\hat{m} = m_0 + \delta\hat{m}$. Since the bare mass m_0 is an undetermined parameter of the theory, the only nonarbitrary value that can be assigned to it is zero, unless it can be attributed to a physical mechanism omitted from the field theory. The dynamical mass δm is in principle determined by the theory, but is found to be infinite in the absence of a renormalization procedure equivalent to introducing an arbitrary cutoff for divergent sums or integrals over relativistic momentum transfer [121].

10.1.2 Classical renormalization and the Dirac equation

The mass-density $\psi^\dagger(\mathbf{x}, t)\gamma^0\hat{m}\psi(\mathbf{x}, t)$ defines an invariant mass integral for the Dirac field,

$$M = \int \psi^\dagger(\mathbf{x}, t)\gamma^0\hat{m}\psi(\mathbf{x}, t)\, d^3\mathbf{x}. \tag{10.4}$$

In quantized theory, this is an operator in the fermion field algebra. Assuming $m_0 = 0$, the mean value $\langle 0|M|0\rangle$ vanishes in the reference vacuum state because all momenta and currents cancel out. In a single-electron state $|a\rangle = a_a^\dagger|0\rangle$, a self-energy (more precisely, self-mass) is defined by $\delta mc^2 = \langle a|Mc^2|a\rangle = \langle a|\int d^3\mathbf{x}\psi^\dagger\gamma^0(-e\not{A})\psi|a\rangle$. Only helicity-breaking virtual transitions can contribute to this electromagnetic self-mass.

Because the Dirac field equation is inherently nondiagonal in the chiral representation, any eigenstate that diagonalizes the operator $\hat{m}$ must be represented by a 4-component Dirac spinor, constructed from two Pauli spinors of opposite helicity. Perturbation theory sums or integrals over momentum-transfer, due to matrix elements of $\hat{m}$, are infinite without some cutoff. Following the logic of quantized QED [121], it is assumed that a covariant cutoff can be defined such that the eigenvalues of this matrix are mc^2, $-mc^2$ for dressed physical electrons of positive and negative energy, respectively. The corresponding classical renormalization replaces an indeterminate mass by its physical value, transforming the spinor representation space to be consistent with this value. A cutoff parameter is to be determined such that the integral defined by Eq. (10.4) reproduces the physical dressed mass m when evaluated using a Dirac spinor field defined for this mass. In the presence of an external Maxwell field, the field equation with diagonalized self-mass is just the Dirac equation for a single dressed electron.

The field equation for a free dressed electron is $(\not{p} - \hat{m})\psi = 0$, in units such that $\hbar = c = 1$. The space-time Green function $G_m(x_1, x_0)$ for this field equation is defined as the solution of

$$(\not{p} - m_0 - \delta\hat{m})G_m(x_1, x_0) = i\delta^{(4)}(x_1, x_0)$$

that vanishes when $t_1 < t_0$. If $G_0(x_1, x_0)$ is defined in the same way, without $\delta\hat{m}$, this implies a Lippmann–Schwinger equation

$$G_m(x_1, x_0) = G_0(x_1, x_0) - i \int d^4y G_0(x_1, y)\delta\hat{m}(y)G_m(y, x_0).$$

In momentum space this takes the form

$$\frac{i}{\not{p} - \hat{m}} = \frac{i}{\not{p} - m_0}(-i\delta\hat{m})\frac{i}{\not{p} - \hat{m}}, \tag{10.5}$$

which produces an infinite sum of Feynman diagrams [120] when expanded by iteration. Individual terms in this expansion are modified by field quantization.

10.2 Symmetry and Noether's theorem

In the formalism of either classical or quantum field theory, symmetries of the Lagrangian density $\mathcal{L}$ or of the action integral $S = \int \mathcal{L} d^4x$ play a vital role in establishing global conservation laws. In classical field theory, Noether's theorem [304, 93, 336] shows that each symmetry of the action integral implies a conservation law. In quantum field theory, such symmetries result in Ward identities [417, 395] for time-ordered products and expectation values. In this context, a symmetry of the formalism is any transformation of the fields and space-time coordinates that leaves the action integral unchanged.

Noether's theorem will be proved here for a classical relativistic theory defined by a generic field ϕ, which may have spinor or tensor indices. The Lagrangian density $\mathcal{L}(\phi, \partial_\mu\phi)$ is assumed to be Lorentz invariant and to depend only on scalar forms defined by spinor or tensor fields. It is assumed that coordinate displacements are described by Jacobi's theorem $\delta(d^4x) = d^4x\, \partial_\mu \delta x^\mu$. The most general variation of the action integral, evaluated over a closed space-time region Ω, is

$$\delta S = \int_\Omega d^4x(\mathcal{L}\partial_\mu\delta x^\mu + \delta\mathcal{L}).$$

Any variation of ϕ takes the form $\delta\phi = \delta_0\phi + \partial_\mu\phi\delta x^\mu$, where δ_0 omits coordinate variations. The full variation $\delta\mathcal{L}$ is

$$\delta\mathcal{L} = (\partial_\mu\mathcal{L})\delta x^\mu + \frac{\partial\mathcal{L}}{\partial\phi}\delta_0\phi + \frac{\partial\mathcal{L}}{\partial(\partial_\mu\phi)}\partial_\mu\delta_0\phi,$$

using $\delta_0\partial_\mu\phi = \partial_\mu\delta_0\phi$. As in integration by parts, a generalized gradient term can be extracted such that

$$\delta\mathcal{L} = (\partial_\mu\mathcal{L})\delta x^\mu + \partial_\mu\left[\frac{\partial\mathcal{L}}{\partial(\partial_\mu\phi)}\delta_0\phi\right] + \left[\frac{\partial\mathcal{L}}{\partial\phi} - \partial_\mu\frac{\partial\mathcal{L}}{\partial(\partial_\mu\phi)}\right]\delta_0\phi.$$

The last term here vanishes when ϕ satisfies the field equations. Using $\mathcal{L}\partial_\mu\delta x^\mu + (\partial_\mu\mathcal{L})\delta x^\mu = \partial_\mu(\mathcal{L}\delta x^\mu)$, the total variation of the action about a field solution takes the form

$$\delta S = -\int_\Omega d^4x\,\partial_\mu J^\mu,$$

where J^μ is the infinitesimal 4-current density

$$-\mathcal{L}\delta x^\mu - \frac{\partial\mathcal{L}}{\partial(\partial_\mu\phi)}\delta_0\phi.$$

Because the assumed hypervolume can be reduced to an infinitesimal, stationary or invariant action implies the local form of Noether's theorem, $\partial_\mu J^\mu = 0$, an equation of continuity in space-time for the generalized current density determined by the field ϕ.

If the variations of both coordinates and field are determined by parameters denoted by α^q, then

$$\delta x^\mu = \frac{\delta x^\mu}{\delta\alpha^q}\delta\alpha^q, \qquad \delta\phi = \frac{\delta\phi}{\delta\alpha^q}\delta\alpha^q.$$

Functional derivatives are indicated here because α^q may be a parametric field, as for example a Lagrange multiplier field. Variations of α^q determine $J^\mu = J^\mu_q d\alpha^q$, where

$$J^\mu_q = -\mathcal{L}\frac{\delta x^\mu}{\delta\alpha^q} - \frac{\partial\mathcal{L}}{\partial(\partial_\mu\phi)}\frac{\delta_0\phi}{\delta\alpha^q} = \left\{\frac{\partial\mathcal{L}}{\partial(\partial_\mu\phi)}\partial_\nu\phi - g^\mu_\nu\mathcal{L}\right\}\frac{\delta x^\nu}{\delta\alpha^q} - \frac{\partial\mathcal{L}}{\partial(\partial_\mu\phi)}\frac{\delta\phi}{\delta\alpha^q}.$$

Green's theorem for the action integral in a hypercylinder enclosed by space-like end surfaces implies that the difference between integrals of J^0 over the forward and backward end 3-surfaces is compensated by the flux $\mathbf{J}$ integrated over the cylinder walls. End surfaces can be defined by a progressive time parameter τ such that $\tau_1 \le \tau \le \tau_2$. If ϕ and its normal gradient vanish on the wall surface for sufficiently large effective radius R, integrals of J^0 over any such nested τ-surface must be equal. This is the integral Noether theorem. If α^q is arbitrary, the integral of J^0_q over any such 3-surface defines a constant of motion for each independent index q.

10.2.1 Examples of conservation laws

Consider an infinitesimal displacement δa^μ in space-time. The action is invariant if for $x' = x + \delta a$, $\phi'(x') = \phi(x)$ for a field ϕ, where the displaced field function is defined by $\phi'(x) = \phi(x) + \delta_0\phi$. These definitions are equivalent to

$$\delta\phi(x) = 0; \qquad \delta x^\mu = \delta a^\mu.$$

The conserved current density determined by Noether's theorem is

$$J^\mu = \left\{ \frac{\partial \mathcal{L}}{\partial(\partial_\mu \phi)} \partial_\nu \phi - g^\mu_\nu \mathcal{L} \right\} \delta a^\nu = T^\mu_\nu \delta a^\nu,$$

defining the *energy–momentum* tensor T^μ_ν. Noether's theorem determines an energy–momentum 4-vector for the field ϕ, conserved if the action integral is 4-translation invariant,

$$P_\nu = \int d^3\mathbf{x}\, T^0_\nu = (E, -c\mathbf{P}),$$

where

$$E = \int d^3\mathbf{x} \left(\frac{\partial \mathcal{L}}{\partial \dot{\phi}} \dot{\phi} - \mathcal{L} \right)$$

$$\mathbf{P} = -\int d^3\mathbf{x} \frac{\partial \mathcal{L}}{\partial \dot{\phi}} \nabla \phi.$$

These formulas are valid for the nonrelativistic one-electron Schrödinger equation. The Lagrangian density is

$$\mathcal{L} = \psi^*(i\hbar\dot{\psi} - \mathcal{H}\psi).$$

Then $\partial\mathcal{L}/\partial\dot{\psi} = i\hbar\psi^*$ implies the standard formulas

$$E = \int d^3\mathbf{x}\, (i\hbar\psi^*\dot{\psi} - \mathcal{L}) = \int d^3\mathbf{x}\, \psi^*\mathcal{H}\psi$$

$$\mathbf{P} = \int d^3\mathbf{x}\, \psi^*(-i\hbar\nabla)\psi.$$

Under a Lorentz transformation Λ, a covariant field $\phi_a(x)$ transforms according to

$$\phi'_a(x') = D_{ab}(\Lambda)\phi_b(\Lambda^{-1}x').$$

For an infinitesimal transformation, the irreducible representation matrix is $D_{ab} = \delta_{ab} + \delta D_{ab}$, and the coordinate transformation returning to the original coordinate value is $(\Lambda^{-1})^\mu_\nu x'^\nu = x'^\mu - \lambda^\mu_\nu x'^\nu = x^\mu$. Defining the transformation with this reverse step makes $\delta x^\mu = 0$, while the functional form of the field changes according to

$$\delta_0\phi_a(x) = \delta D_{ab}\phi_b(x) - \lambda^\mu_\nu x^\nu \partial_\mu \phi_a(x).$$

The conserved current defined by Noether's theorem is

$$J^\mu = -\frac{\partial \mathcal{L}}{\partial(\partial_\mu \phi_a)} \left(\delta D_{ab}\phi_b(x) - \lambda^\mu_\nu x^\nu \partial_\mu \phi_a(x) \right).$$

In the case of a scalar field, the irreducible matrix D is a unit matrix, and drops out of J^μ. For rotation through an angle $\delta\theta_k$ about the Cartesian axis $\hat{e}_k$, the rotational submatrix of the Lorentz matrix is given by $\lambda^i_j x^j = \delta\theta_k \epsilon^{kij} x^j$, where ϵ^{ijk} is the totally antisymmetric Levi–Civita tensor. For the one-electron Schrödinger field ψ, Noether's theorem defines three conserved components of a spatial axial vector,

$$L^i = -\int d^3\mathbf{x}\, \epsilon^{ijk} \frac{\partial \mathcal{L}}{\partial \dot{\psi}} x^j \nabla_k \psi,$$

which is just the orbital angular momentum vector

$$\mathbf{L} = \int d^3\mathbf{x}\, \psi^* \mathbf{r} \times (-i\hbar \boldsymbol{\nabla})\psi.$$

For the Dirac bispinor, the irreducible representation matrix D_{ab} for each helicity component is a Pauli spin matrix $\boldsymbol{\sigma}$ multiplied by $\hbar/2$. Then

$$\delta D_{ab}\psi_b = \frac{1}{2}\hbar\delta\theta_k \sigma^k_{ab}\psi_b.$$

Given

$$\partial\mathcal{L}/\partial(\partial_\mu\psi) = i\hbar c \psi^\dagger \gamma^0 \gamma^\mu,$$

and $\gamma^0\gamma^0 = I$, the conserved angular momentum is

$$\mathbf{J} = \int d^3\mathbf{x}\, \psi^\dagger \left\{ \frac{1}{2}\hbar\boldsymbol{\sigma} + \mathbf{r} \times (-i\hbar\boldsymbol{\nabla}) \right\} \psi = \mathbf{S} + \mathbf{L}.$$

S is the elementary spin of the Dirac electron.

10.3 Gauge invariance

In classical electrodynamics, the field equations for the Maxwell field A_μ depend only on the antisymmetric tensor $F_{\mu\nu}$, which is invariant under a gauge transformation $A_\mu \to A_\mu + \hbar c \partial_\mu \chi(x)$, where χ is an arbitrary scalar field in space-time. Thus the vector field A_μ is not completely determined by the theory. It is customary to impose an auxiliary gauge condition, such as $\partial_\mu A^\mu = 0$, in order to simplify the field equations. In the presence of an externally determined electric current density 4-vector j^μ, the Maxwell Lagrangian density is

$$\mathcal{L}_A = -\frac{1}{16\pi} F^{\mu\nu} F_{\mu\nu} - \frac{1}{c} j^\mu A_\mu.$$

The Euler–Lagrange equations are

$$\partial_\mu \frac{\partial \mathcal{L}_A}{\partial(\partial_\mu A_\nu)} - \frac{\partial \mathcal{L}_A}{\partial A_\nu} = 0,$$

which implies the covariant inhomogeneous Maxwell equations

$$\partial_\mu F^{\mu\nu} = (4\pi/c)j^\nu.$$

Because $F^{\mu\nu}$ is antisymmetric, the symmetrical derivative $\partial_\mu\partial_\nu F^{\mu\nu}$ must vanish. This requires j^ν to satisfy the equation of continuity, $\partial_\nu j^\nu = 0$, which implies charge conservation in an enclosed volume if net current flow vanishes across its spatial boundary.

Gauge covariance of the classical theory is due to the invariance of the field tensor $F_{\mu\nu}$ under the local gauge transformation

$$A_\mu(x) \to A_\mu(x) + \hbar c \partial_\mu \chi(x),$$

on the assumption that the current density j^μ is not affected by this transformation. However, the Lagrangian density $\mathcal{L}_A$ is not invariant, since the coupling term is modified such that

$$\mathcal{L}_A \to \mathcal{L}_A - \hbar j^\mu \partial_\mu \chi(x).$$

Thus the equations of motion are gauge invariant, but the action integral is not.

10.3.1 Classical electrodynamics as a gauge theory

Analysis of the "classical" Dirac theory shows a similar inconsistency under local phase transformations, such that $\psi(x) \to e^{ie\chi(x)}\psi(x)$, corresponding to the local infinitesimal transformation, for $\chi \to 0$,

$$\delta\psi(x) = ie\chi(x)\psi(x).$$

For the Dirac field in an externally determined Maxwell field, the Lagrangian density including a renormalized mass term is

$$\mathcal{L}_D = i\hbar c\psi^\dagger\gamma^0\{\gamma^\mu D_\mu + im\}\psi.$$

The field interaction is expressed in terms of the covariant derivative

$$D_\mu = \partial_\mu + i(-e/\hbar c)A_\mu,$$

such that the Euler–Lagrange equation for the Dirac field is

$$i\hbar c\gamma^\mu D_\mu\psi = mc^2\psi.$$

This is gauge-covariant only if the gauge transformation of A_μ and the phase transformation of ψ are combined, such that

$$A_\mu \to A_\mu + \hbar c \partial_\mu \chi(x); \qquad \psi \to e^{ie\chi(x)}\psi.$$

Because

$$e^{-ie\chi(x)} D_\mu \psi \to \left\{ \partial_\mu + ie(\partial_\mu \chi) - \frac{ie}{\hbar c}(A_\mu + \hbar c \partial_\mu \chi) \right\} \psi = D_\mu \psi,$$

the Euler–Lagrange equation for ψ is covariant under this generalized local gauge transformation. The electrodynamic field action, expressed in terms of the gauge invariant tensor $F_{\mu\nu}$ and the covariant derivative $D_\mu \psi$, is gauge invariant. This introduces a scalar gauge function $\chi(x)$ that is inherently unobservable. Gauge invariance completely determines the interaction between the Dirac and Maxwell fields.

10.3.2 Noether's theorem for gauge symmetry

For a local infinitesimal gauge transformation about a solution of the field equations,

$$\delta \mathcal{L} = \partial_\mu \left[\frac{\partial \mathcal{L}}{\partial(\partial_\mu A_\nu)} \delta A_\nu + \frac{\partial \mathcal{L}}{\partial(\partial_\mu \psi)} \delta \psi \right] = \partial_\mu \left(-\frac{\hbar c}{4\pi} F^{\mu\nu} \partial_\nu \chi + \hbar j^\mu \chi \right) = 0,$$

defining the Dirac electric current density

$$j^\mu(x) = -ec\psi^\dagger(x)\gamma^0\gamma^\mu\psi(x).$$

Because antisymmetry of $F^{\mu\nu}$ implies that $F^{\mu\nu}\partial_\mu\partial_\nu\chi \equiv 0$,

$$(\partial_\mu j^\mu)\chi(x) = \left(\frac{c}{4\pi} \partial_\nu F^{\nu\mu} - j^\mu \right) \partial_\mu \chi(x) = 0.$$

Thus, since $\chi(x)$ is arbitrary, Noether's theorem implies $\partial_\mu j^\mu = 0$, as required for consistency of the field equations. While j^μ is invariant under a local *phase* transformation, local *gauge* invariance requires an interaction described by the covariant derivative, implying a gauge-covariant Maxwell field. This has profound consequences in the generalization to electroweak theory [340, 217], since renormalizability of the quantum field theory follows from generalized gauge invariance [404, 405]. Noether's theorem implies that gauge fields and the corresponding conserved current densities must exist in any renormalizable quantum theory that can describe the weak interactions responsible for beta-decay processes.

10.3.3 Nonabelian gauge symmetries

In electrodynamics, the gauge function $\chi(x)$ is a real scalar function in space-time, and the phase transformation of the fermion field takes the form $\psi(x) \to e^{ie\chi(x)}\psi(x)$. This is a multiplicative unitary transformation, characterized by $U(1)$ in group theory [217]. Because the representation matrix is simply the phase factor $e^{ie\chi(x)}$, any two such factors commute, defining an abelian group. As originally shown by Yang and Mills [435], it is possible to construct field theories with gauge symmetries in which $U(1)$ phase factors are replaced by noncommuting unitary representation matrices of a nonabelian continuous group. For example, the group $SU(2)$ is represented by 2×2 unitary matrices $\exp(ie\boldsymbol{\chi}(x) \cdot \boldsymbol{\tau})$, where the matrices $\boldsymbol{\tau}$ are identical to the Pauli matrices $\boldsymbol{\sigma}$.

Electroweak theory (EWT) [73, 340, 217] makes use of the gauge group $U(1) \times SU(2)$. The representation matrices are of the form

$$\exp(ie\chi(x)\tau_0 + ie'\boldsymbol{\chi}(x) \cdot \boldsymbol{\tau}),$$

where τ_0 is the unit matrix. This is a product of $U(1)$ factors and 2×2 $SU(2)$ matrices. The fermion field ψ becomes an array consisting of two spinors, each carrying an additional "weak isospin" index I_3. In EWT this index distinguishes members of isospin doublets: electrons and neutrinos or down quarks and up quarks.

A local unitary transformation of a fermion field ψ is defined by

$$\psi \to U\psi; \qquad \psi^\dagger \to \psi^\dagger U^{-1},$$

where U commutes with the Dirac matrices γ^μ. Interaction with a gauge field E_μ is described by a covariant derivative D_μ in the fermion Lagrangian density

$$\mathcal{L} = i\hbar c\psi^\dagger\gamma^0\gamma^\mu D_\mu\psi = i\hbar c\psi^\dagger\gamma^0\gamma^\mu(\partial_\mu + E_\mu)\psi.$$

If the gauge field transforms according to

$$E_\mu \to UE_\mu U^{-1} - (\partial_\mu U)U^{-1}, \tag{10.6}$$

this ensures gauge invariance of $\mathcal{L}$,

$$\mathcal{L} \to i\hbar c\psi^\dagger\gamma^0\gamma^\mu U^{-1}[U\partial_\mu + (\partial_\mu U) + UE_\mu - (\partial_\mu U)]\psi = \mathcal{L}.$$

For the abelian phase factor of the $U(1)$ group, $U = \exp\{ie\chi(x)\}$ and $E_\mu = -(ie/\hbar c)A_\mu$, which produces the usual gauge transformation $A_\mu \to A_\mu + \hbar c\partial_\mu\chi(x)$. In electroweak theory, a $U(1)$ gauge field B_μ is defined such that $U = \exp\{-\frac{1}{2}ig'y\chi(x)\}$ and $E_\mu = (ig'y/2\hbar c)B_\mu$. The gauge transformation is $B_\mu \to B_\mu + \hbar c\partial_\mu\chi(x)$.

The transformation matrix for the nonabelian group $SU(2)$ is

$$U = \exp\left\{-i\frac{g}{2}\chi(x)\cdot\boldsymbol{\tau}\right\}.$$

The $SU(2)$ gauge field $\mathbf{W}_\mu$ has three components, corresponding to the isospin vector of matrices $\boldsymbol{\tau}$, with no relationship to the coordinate space $ct, \mathbf{x}$. By implication, the fermion field ψ is a set of spinors, one for each value of the isospin index. The covariant derivative

$$D_\mu = \partial_\mu + i\frac{g}{2\hbar c}\boldsymbol{\tau}\cdot\mathbf{W}_\mu$$

ensures that the fermion Lagrangian density $\mathcal{L} = i\hbar c\psi^\dagger\gamma^0\gamma^\mu D_\mu\psi$ is invariant if the gauge field transforms according to Eq. (10.6). This Lagrangian density describes physical processes associated with beta decay, transitions that convert a neutrino into a lepton or convert a down quark into an up quark (or a neutron into a proton), coupling such transitions in second order.

Considering only the $SU(2)$ field interaction, the Euler–Lagrange equation for the fermion field is

$$i\hbar c\gamma^\mu\partial_\mu\psi = \frac{1}{2}g\gamma^\mu\boldsymbol{\tau}\cdot\mathbf{W}_\mu\psi.$$

If there is no independent external gauge field, this defines an incremental fermion mass operator $\delta\hat{m}_W c^2 = \frac{1}{2}g\gamma^\mu\boldsymbol{\tau}\cdot\mathbf{W}_\mu$ such that

$$i\hbar c\gamma^\mu\partial_\mu\psi = \delta\hat{m}_W c^2\psi.$$

As in the case of the electromagnetic self-mass, the implied dynamical mass increment is infinite unless perturbation-theory sums are truncated by a renormalization cutoff procedure. In analogy to electrodynamics, each fermion field acquires an incremental dynamical mass through interaction with the gauge field. This implies in electroweak theory that neutrinos must acquire such a dynamical mass from their interaction with the $SU(2)$ gauge field. For a renormalized Dirac fermion in an externally determined $SU(2)$ gauge field, the Lagrangian density is

$$\mathcal{L}_D = i\hbar c\psi^\dagger\gamma^0\{\gamma^\mu\partial_\mu + im\}\psi - \frac{1}{c}\mathbf{j}_W^\mu\cdot\mathbf{W}_\mu,$$

where

$$\mathbf{j}_W^\mu = \frac{1}{2}gc\psi^\dagger\gamma^0\gamma^\mu\boldsymbol{\tau}\psi.$$

This defines the fermion contribution to an isovector gauge current density. Although the Euler–Lagrange equation is gauge covariant by construction, this fermion gauge current is not invariant, because the matrix $\boldsymbol{\tau}$ does not commute with the $SU(2)$ unitary transformation matrices. It will be shown below that the conserved Noether current density augments $\mathbf{j}_W^\mu$ by a component determined by the gauge field. The nonabelian gauge field carries a nonvanishing isospin, which is implicit in the isovector notation used here for $\mathbf{W}_\mu$. It interacts with itself through the Noether current density.

Transformation properties of $\mathbf{W}_\mu$ can be derived by considering an infinitesimal $SU(2)$ gauge transformation, $U = I - \frac{1}{2}ig\boldsymbol{\chi}(x)\cdot\boldsymbol{\tau}$, for $\boldsymbol{\chi}(x) \to 0$. The corresponding infinitesimal transformation of the fermion field is

$$\psi \to \psi - \frac{1}{2}ig\boldsymbol{\chi}(x)\cdot\boldsymbol{\tau}\psi.$$

Because the matrix $\boldsymbol{\chi}(x)\cdot\boldsymbol{\tau}$ can convert a neutrino into an electron at any point in space-time, a physically correct theory must couple such a transition with a field that carries a compensating electric charge. Thus local $SU(2)$ gauge invariance implies existence of the gauge field $\mathbf{W}_\mu$ with components that carry electric charge. The field equations for $\mathbf{W}_\mu$ contain a source term with off-diagonal elements that represent charge transitions. Retaining only first-order terms in $\boldsymbol{\chi}(x)$, Eq. (10.6) takes the form [435, 340]

$$\begin{aligned}\boldsymbol{\tau}\cdot\mathbf{W}_\mu \to \boldsymbol{\tau}\cdot\mathbf{W}_\mu + \frac{1}{2}ig(\boldsymbol{\tau}\cdot\mathbf{W}_\mu\boldsymbol{\chi}\cdot\boldsymbol{\tau} - \boldsymbol{\chi}\cdot\boldsymbol{\tau}\boldsymbol{\tau}\cdot\mathbf{W}_\mu) + \hbar c\boldsymbol{\tau}\cdot\partial_\mu\boldsymbol{\chi}\\ = \boldsymbol{\tau}\cdot(\mathbf{W}_\mu - g\mathbf{W}_\mu\times\boldsymbol{\chi} + \hbar c\partial_\mu\boldsymbol{\chi}).\end{aligned}$$

This derivation uses an algebraic identity valid for Pauli matrices and any two vectors $\mathbf{A}$, $\mathbf{B}$,

$$(\boldsymbol{\tau}\cdot\mathbf{A})(\boldsymbol{\tau}\cdot\mathbf{B}) = \mathbf{A}\cdot\mathbf{B} + i\boldsymbol{\tau}\cdot\mathbf{A}\times\mathbf{B}.$$

Since the components of $\boldsymbol{\tau}$ are linearly independent, the implied infinitesimal gauge transformation is

$$\mathbf{W}_\mu \to \mathbf{W}_\mu - g\mathbf{W}_\mu\times\boldsymbol{\chi} + \hbar c\partial_\mu\boldsymbol{\chi}.$$

The term proportional to the coupling constant g is typical of a nonabelian gauge group [422].

In the presence of an externally determined fermion gauge current, the Lagrangian density for the $SU(2)$ gauge field is [435]

$$\mathcal{L}_W = -\frac{1}{16\pi}\mathbf{W}^{\mu\nu}\cdot\mathbf{W}_{\mu\nu} - \frac{1}{c}\mathbf{j}_W^\mu\cdot\mathbf{W}_\mu,$$

expressed in terms of an isovector field defined by

$$\mathbf{W}_{\mu\nu} = \partial_\mu\mathbf{W}_\nu - \partial_\nu\mathbf{W}_\mu - \frac{g}{\hbar c}\mathbf{W}_\mu\times\mathbf{W}_\nu.$$

The first term in $\mathcal{L}_W$ is gauge invariant by construction [340, 435]. The second term is not, because of the gauge modification of $\mathbf{W}_\mu$. The implied Euler–Lagrange equations for the gauge field [435] are

$$\partial_\mu \mathbf{W}^{\mu\nu} = \frac{g}{\hbar c}\mathbf{W}_\mu \times \mathbf{W}^{\mu\nu} + \frac{4\pi}{c}\mathbf{j}_W^\nu = \frac{4\pi}{c}\mathbf{J}_W^\nu,$$

defining the total isovector current density $\mathbf{J}_W^\nu$. It will be shown here that this is the conserved Noether current density for the $SU(2)$ theory. In contrast with electrodynamics, a nonabelian gauge field interacts with itself through its contribution to this conserved current density.

10.3.4 Gauge invariance of the SU(2) field theory

The Lagrangian density for a massless fermion field interacting with the $SU(2)$ gauge field is

$$\mathcal{L} = -\frac{1}{16\pi}\mathbf{W}^{\mu\nu}\cdot\mathbf{W}_{\mu\nu} + i\hbar c\psi^\dagger\gamma^0\gamma^\mu D_\mu\psi,$$

expressed in terms of the covariant derivative

$$D_\mu = \partial_\mu + i\frac{g}{2\hbar c}\boldsymbol{\tau}\cdot\mathbf{W}_\mu.$$

Some of the algebraic complexity of the proof of gauge invariance and of derivation of the conserved Noether gauge current for this Lagrangian density can be simplified by considering general variations of the fermion and gauge fields about a solution of the field equations. The logic of this derivation is unchanged if a fermion mass is included. Stationary action implies that

$$\int d^4x\,\delta\mathcal{L} = \int d^4x\left\{-\frac{1}{8\pi}\mathbf{W}^{\mu\nu}\cdot\delta\mathbf{W}_{\mu\nu} + i\hbar c(\delta\psi^\dagger)\gamma^0\gamma^\mu D_\mu\psi\right.$$
$$\left. + i\hbar c\psi^\dagger\gamma^0\gamma^\mu\delta(D_\mu\psi)\right\} = 0.$$

The fermion field equation $i\hbar\gamma^\mu D_\mu\psi = 0$ is implied by independent variation of $\psi^\dagger$. The Euler–Lagrange equations for the gauge field follow from

$$\partial_\mu\frac{\partial\mathcal{L}}{\partial\mathbf{W}_{\tau\sigma}}\cdot\frac{\partial\mathbf{W}_{\tau\sigma}}{\partial(\partial_\mu\mathbf{W}_\nu)} = \frac{\partial\mathcal{L}}{\partial\mathbf{W}_{\tau\sigma}}\cdot\frac{\partial\mathbf{W}_{\tau\sigma}}{\partial\mathbf{W}_\nu} + \frac{\partial\mathcal{L}}{\partial(D_\mu\psi)}\frac{\partial(D_\mu\psi)}{\partial\mathbf{W}_\nu}.$$

The individual terms here are

$$\frac{\partial\mathcal{L}}{\partial\mathbf{W}_{\tau\sigma}}\cdot\frac{\partial\mathbf{W}_{\tau\sigma}}{\partial(\partial_\mu\mathbf{W}_\nu)} = -\frac{1}{4\pi}\mathbf{W}^{\mu\nu},$$

$$\frac{\partial\mathcal{L}}{\partial\mathbf{W}_{\tau\sigma}}\cdot\frac{\partial\mathbf{W}_{\tau\sigma}}{\partial\mathbf{W}_\nu} = -\frac{g}{4\pi\hbar c}\mathbf{W}_\mu\times\mathbf{W}^{\mu\nu},$$

$$\frac{\partial\mathcal{L}}{\partial(D_\mu\psi)}\frac{\partial(D_\mu\psi)}{\partial\mathbf{W}_\nu} = -\frac{1}{2}g\psi^\dagger\gamma^0\gamma^\nu\boldsymbol{\tau}\psi,$$

implying the gauge field equations

$$\partial_\mu \mathbf{W}^{\mu\nu} = \frac{4\pi}{c}\left(\frac{g}{4\pi\hbar}\mathbf{W}_\mu \times \mathbf{W}^{\mu\nu} + \frac{1}{2}gc\psi^\dagger\gamma^0\gamma^\nu\boldsymbol{\tau}\psi\right) = \frac{4\pi}{c}\mathbf{J}_W^\nu.$$

Because $\mathbf{W}^{\mu\nu}$ is an antisymmetric Lorentz tensor, the total isovector current density must satisfy $\partial_\nu \mathbf{J}_W^\nu = 0$.

Field variations driven by the infinitesimal local gauge transformation

$$\delta\psi = -\frac{1}{2}ig\boldsymbol{\tau}\cdot\boldsymbol{\chi}(x)\psi,$$
$$\delta\mathbf{W}_\mu = -g\mathbf{W}_\mu \times \boldsymbol{\chi}(x) + \hbar c\partial_\mu\boldsymbol{\chi}(x),$$

where $\boldsymbol{\chi}(x) \to 0$, determine the conserved Noether current density implied by gauge invariance. D_μ and the gauge field transformation are defined so that the fermion part of $\mathcal{L}$ is gauge invariant. To evaluate the variation of the gauge field part, direct substitution of $\delta\mathbf{W}_\mu$ into the definition of $\mathbf{W}_{\mu\nu}$ gives

$$\begin{aligned}\delta\mathbf{W}_{\mu\nu} &= \partial_\mu\delta\mathbf{W}_\nu - \partial_\nu\delta\mathbf{W}_\mu - \frac{g}{\hbar c}(\mathbf{W}_\mu \times \delta\mathbf{W}_\nu - \mathbf{W}_\nu \times \delta\mathbf{W}_\mu)\\ &= -g\left(\partial_\mu\mathbf{W}_\nu - \partial_\nu\mathbf{W}_\mu - \frac{g}{\hbar c}\mathbf{W}_\mu \times \mathbf{W}_\nu\right) \times \boldsymbol{\chi}\\ &= -g\mathbf{W}_{\mu\nu} \times \boldsymbol{\chi},\end{aligned}$$

using the vector identity $\mathbf{a}\times(\mathbf{b}\times\mathbf{c}) - \mathbf{b}\times(\mathbf{a}\times\mathbf{c}) = (\mathbf{a}\times\mathbf{b})\times\mathbf{c}$. Gauge invariance follows from $\mathbf{W}^{\mu\nu}\cdot\mathbf{W}_{\mu\nu}\times\boldsymbol{\chi} = \mathbf{W}^{\mu\nu}\times\mathbf{W}_{\mu\nu}\cdot\boldsymbol{\chi} \equiv 0$.

For fixed coordinates, invariance of the action integral for gauge variations about $SU(2)$ field solutions implies

$$\begin{aligned}\delta\mathcal{L} &= \partial_\mu\left[\frac{\partial\mathcal{L}}{\partial(\partial_\mu\mathbf{W}_\nu)}\cdot\delta\mathbf{W}_\nu + \frac{\partial\mathcal{L}}{\partial(\partial_\mu\psi)}\delta\psi\right]\\ &= \partial_\mu\left[\frac{1}{4\pi}\mathbf{W}^{\mu\nu}\cdot(g\mathbf{W}_\nu\times\boldsymbol{\chi} - \hbar c\partial_\nu\boldsymbol{\chi}) + \frac{1}{2}\hbar gc\psi^\dagger\gamma^0\gamma^\mu\boldsymbol{\tau}\psi\cdot\boldsymbol{\chi}\right]\\ &= \partial_\mu\left(\hbar\mathbf{J}_W^\mu\cdot\boldsymbol{\chi} - \frac{\hbar c}{4\pi}\mathbf{W}^{\mu\nu}\cdot\partial_\nu\boldsymbol{\chi}\right) = 0.\end{aligned}$$

Because $\mathbf{W}^{\mu\nu}\cdot\partial_\mu\partial_\nu\boldsymbol{\chi}$ vanishes identically, due to the antisymmetry of $\mathbf{W}^{\mu\nu}$, this implies Noether's theorem in the form

$$(\partial_\mu\mathbf{J}_W^\mu)\cdot\boldsymbol{\chi}(x) = \left(\frac{c}{4\pi}\partial_\nu\mathbf{W}^{\nu\mu} - \mathbf{J}_W^\mu\right)\cdot\partial_\mu\boldsymbol{\chi}(x) = 0.$$

Because $\boldsymbol{\chi}(x)$ is arbitrary, the total isovector current satisfies $\partial_\mu\mathbf{J}_W^\mu = 0$ at all points in space-time so long as the field equations are satisfied.

10.4 Energy and momentum of the coupled fields

10.4.1 Energy and momentum in classical electrodynamics

The energy–momentum tensor derived from Noether's theorem for electrodynamics is

$$T^{\mu}_{\nu} = \frac{\partial \mathcal{L}}{\partial(\partial_\mu A_\lambda)}\partial_\nu A_\lambda + \frac{\partial \mathcal{L}}{\partial(\partial_\mu \psi)}\partial_\nu \psi - g^{\mu}_{\nu}\mathcal{L}.$$

The Lorentz momentum 4-vector is

$$P_\nu = \int d^3x\, T^0_\nu = (E, -c\mathbf{P}),$$

in any inertial frame, where $T^0_\nu = T^0_\nu(\psi) + T^0_\nu(A)$ such that

$$T^0_\nu(\psi) = i\hbar c\psi^\dagger(\partial_\nu\psi - \delta_{0\nu}\gamma^0\gamma^\lambda D_\lambda\psi),$$

$$T^0_\nu(A) = -\frac{1}{4\pi}F^{0\lambda}\partial_\nu A_\lambda + \delta_{0\nu}\frac{1}{16\pi}F^{\sigma\lambda}F_{\sigma\lambda}.$$

Energy–momentum conservation is expressed by $\partial^\nu T^0_\nu = 0$ for a closed system. If T^μ_ν were a symmetric tensor (when converted to $T^{\mu\nu}$), this would be assured because $\partial_\mu T^\mu_\nu = 0$ by construction. Since the gauge field part of the tensor deduced from Noether's theorem is not symmetric, this requires special consideration, as discussed below. A symmetric energy–momentum tensor is required for any eventual unification of quantum field theory and general relativity [422]. The fermion field energy and momentum are

$$E(\psi) = \int d^3x\, \psi^\dagger \mathcal{H}\psi, \qquad \mathbf{P}(\psi) = \int d^3x\, \psi^\dagger(-i\hbar\nabla)\psi,$$

where, for a massless electron of negative helicity (positive energy),

$$\mathcal{H} = -i\hbar(c\gamma^0\gamma^\lambda D_\lambda - \partial_t) = -c\boldsymbol{\sigma}\cdot\left(\mathbf{p} + \frac{e}{c}\mathbf{A}\right) - e\phi.$$

For the Maxwell field, the energy–momentum tensor $T^\mu_\nu(A)$ derived from Noether's theorem is unsymmetric, and not gauge invariant, in contrast to the symmetric stress tensor derived directly from Maxwell's equations [318]. Consider the symmetric tensor $\Theta = T + \Delta T$, where

$$\Delta T^\mu_\nu = \frac{1}{4\pi}F^{\mu\lambda}\partial_\lambda A_\nu$$

is defined such that

$$\partial_\mu \Delta T^\mu_\nu = \frac{1}{4\pi}(\partial_\mu F^{\mu\lambda})\partial_\lambda A_\nu + \frac{1}{4\pi}F^{\mu\lambda}\partial_\mu\partial_\lambda A_\nu = \frac{1}{c}j^\lambda\partial_\lambda A_\nu,$$

using the antisymmetry of $F^{\mu\lambda}$ and the field equations. For a noninteracting field the residual term vanishes and the resulting symmetrical energy–momentum tensor is

$$\Theta^{\mu}_{\nu}(A) = -\frac{1}{4\pi}F^{\mu\lambda}F_{\nu\lambda} + g^{\mu}_{\nu}\frac{1}{16\pi}F^{\sigma\lambda}F_{\sigma\lambda}.$$

The electric and magnetic field vectors in vacuo are, respectively,

$$\mathcal{E} = -\frac{1}{c}\frac{\partial \mathbf{A}}{\partial t} - \nabla\phi, \qquad \mathcal{B} = \nabla \times \mathbf{A}.$$

Using italic indices for 3-vectors, with $A_\mu = (\phi, -\mathbf{A})$, $\partial_\mu = (\partial/c\partial t, \nabla)$,

$$\mathcal{E}_i = \partial_0 A_i - \partial_i A_0 = F_{0i},$$
$$\mathcal{B}_i = -\epsilon_{ijk}(\partial_j A_k - \partial_k A_j) = -\epsilon_{ijk}F_{jk}.$$

In terms of these vector fields, the Maxwell field Lagrangian density is $(1/8\pi)(\mathcal{E}^2 - \mathcal{B}^2)$, and the field energy and momentum are

$$E(A) = \int d^3x \frac{\mathcal{E}^2 + \mathcal{B}^2}{8\pi}, \qquad \mathbf{P}(A) = \int d^3x \frac{\mathcal{E} \times \mathcal{B}}{4\pi c}.$$

For interacting fields, the Maxwell field energy is not separately conserved. A gauge-covariant derivation follows from the inhomogeneous field equations (Maxwell equations in vacuo),

$$\nabla \times \mathcal{E} + \frac{\partial}{c\partial t}\mathcal{B} = 0, \qquad \nabla \cdot \mathcal{E} = 4\pi\rho,$$
$$\nabla \times \mathcal{B} - \frac{\partial}{c\partial t}\mathcal{E} = \frac{4\pi}{c}\mathbf{j}, \qquad \nabla \cdot \mathcal{B} = 0.$$

The local conservation law follows immediately [318] from

$$\mathcal{B}\cdot(\nabla \times \mathcal{E}) - \mathcal{E}\cdot(\nabla \times \mathcal{B}) = \nabla\cdot(\mathcal{E} \times \mathcal{B}) = -\frac{1}{2}\frac{\partial}{c\partial t}(\mathcal{E}^2 + \mathcal{B}^2) - \frac{4\pi}{c}\mathcal{E}\cdot\mathbf{j}.$$

Expressed in terms of the field energy and momentum densities defined above, this is

$$\frac{\partial}{\partial t}\frac{\mathcal{E}^2 + \mathcal{B}^2}{8\pi} + c^2\nabla\cdot\frac{\mathcal{E} \times \mathcal{B}}{4\pi c} = -\mathcal{E}\cdot\mathbf{j}.$$

When integrated over a finite volume, the divergence term becomes a surface integral of the Poynting vector [318]. Expressed in covariant notation, $\partial^\nu\Theta^0_\nu = -\frac{1}{c}F^{0\nu}j_\nu$, showing the explicit modification of energy–momentum conservation due to the final dissipative term here.

As a model for classical gauge fields, the energy–momentum conservation law can be derived directly in covariant notation. The 4-divergence

$$-\frac{1}{4\pi}\partial^\nu(F^{0\lambda}F_{\nu\lambda}) = -\frac{1}{4\pi}\partial^i(F^{0j}F_{ij}) - \frac{1}{4\pi}\partial^0(F^{0j}F_{0j})$$

can also be written as

$$-\frac{1}{4\pi}(\partial^\nu F^{0\lambda})F_{\nu\lambda} - \frac{1}{4\pi}F^{0\lambda}(\partial^\nu F_{\nu\lambda}) = -\frac{1}{16\pi}\partial^0(F^{\nu\lambda}F_{\nu\lambda}) - \frac{1}{c}F^{0\lambda}j_\lambda,$$

using the field equations and the identity

$$(\partial^\nu F^{0\lambda})F_{\nu\lambda} \equiv \frac{1}{4}\partial^0(F^{\nu\lambda}F_{\nu\lambda}).$$

This establishes, as a consequence of the field equations,

$$-\frac{1}{8\pi}\partial^0(F^{0i}F_{0i} - F^{jk}F_{jk}|_{j<k}) - \frac{1}{4\pi}\partial^i(F^{0j}F_{ij}) = -\frac{1}{c}F^{0i}j_i,$$

verifying the conservation law.

Since local energy and momentum density are well-defined for the classical Maxwell field, respectively

$$\epsilon(x) = \frac{\mathcal{E}^2 + \mathcal{B}^2}{8\pi}, \qquad \pi(x) = \frac{\mathcal{E}\times\mathcal{B}}{4\pi c},$$

the relativistic mass density of the field is defined by

$$\mu(x)^2c^4 = \epsilon(x)^2 - \pi(x)^2c^2,$$

without invoking a specific requirement for a symmetric stress tensor. For a plane-wave radiation field in vacuo,

$$\mu^2c^4 = \frac{(\mathcal{E}^2 - \mathcal{B}^2)^2 + 4(\mathcal{E}\cdot\mathcal{B})^2}{64\pi^2} = 0,$$

because the field vectors are orthogonal and of equal magnitude (in mixed Gaussian units).

10.4.2 Energy and momentum in SU(2) gauge theory

The Lagrangian density for a massless fermion field interacting with the $SU(2)$ gauge field defines the Noether energy–momentum tensor

$$T^\mu_\nu = \frac{\partial\mathcal{L}}{\partial(\partial_\mu\mathbf{W}_\lambda)}\cdot\partial_\nu\mathbf{W}_\lambda + \frac{\partial\mathcal{L}}{\partial(\partial_\mu\psi)}\partial_\nu\psi - g^\mu_\nu\mathcal{L}.$$

The fermion field energy and momentum are

$$E(\psi) = \int d^3x\, \psi^\dagger \mathcal{H}_W \psi, \qquad \mathbf{P}(\psi) = \int d^3x\, \psi^\dagger(-i\hbar\boldsymbol{\nabla})\psi.$$

Using italic indices for spatial 3-vectors, but retaining vector notation for the abstract 3-vector index of the matrices $\boldsymbol{\tau}$ and the gauge field $\mathbf{W}$, the fermion "Hamiltonian" operator is

$$\mathcal{H}_W = -i\hbar(c\gamma^0\gamma^\lambda D_\lambda - \partial_t) = -c\sigma_i\left(p_i - \frac{g}{2c}\boldsymbol{\tau}\cdot\mathbf{W}_i\right) + \frac{1}{2}g\boldsymbol{\tau}\cdot\mathbf{W}_0.$$

The $SU(2)$ energy–momentum tensor can be symmetrized and made gauge invariant by adding an incremental tensor

$$\Delta T^\mu_\nu = \frac{1}{4\pi}\mathbf{W}^{\mu\lambda}\cdot\left(\partial_\lambda\mathbf{W}_\nu + \frac{g}{\hbar c}\mathbf{W}_\nu \times \mathbf{W}_\lambda\right).$$

Because $\partial_\mu \Delta T^\mu_\nu$ does not reduce to terms that vanish even for a noninteracting field, this construction must be verified. The energy and 3-momentum of the gauge field derived from the resulting symmetric energy–momentum tensor Θ^μ_ν are

$$E(W) = -\frac{1}{8\pi}\int d^3x(\mathbf{W}^{0i}\cdot\mathbf{W}_{0i} - \mathbf{W}^{jk}\cdot\mathbf{W}_{jk}|_{j<k}),$$

$$P_i(W) = \frac{1}{4\pi c}\int d^3x\, \mathbf{W}^{0j}\cdot\mathbf{W}_{ij}.$$

The local conservation law for the interacting gauge field can be derived from the covariant field equations, as was done above for the Maxwell field. Using the $SU(2)$ field equations and expanding $(\partial^\nu\mathbf{W}^{0\lambda})\cdot\mathbf{W}_{\nu\lambda}$ as

$$\frac{1}{4}\partial^0(\mathbf{W}^{\nu\lambda}\cdot\mathbf{W}_{\nu\lambda}) + \frac{g}{2\hbar c}[\partial^0(\mathbf{W}^\nu\times\mathbf{W}^\lambda) - 2\partial^\nu(\mathbf{W}^0\times\mathbf{W}^\lambda)]\cdot\mathbf{W}_{\nu\lambda},$$

there are two distinct expansions of the 4-divergence

$$\begin{aligned}
-\frac{1}{4\pi}\partial^\nu(\mathbf{W}^{0\lambda}\cdot\mathbf{W}_{\nu\lambda}) &= -\frac{1}{4\pi}\partial^i(\mathbf{W}^{0j}\cdot\mathbf{W}_{ij}) - \frac{1}{4\pi}\partial^0(\mathbf{W}^{0j}\cdot\mathbf{W}_{0j}) \\
&= -\frac{1}{16\pi}\partial^0(\mathbf{W}^{\nu\lambda}\cdot\mathbf{W}_{\nu\lambda}) - \frac{1}{c}\mathbf{W}^{0\lambda}\cdot\mathbf{J}^W_\lambda \\
&\quad - \frac{g}{8\pi\hbar c}[\partial^0(\mathbf{W}^\nu\times\mathbf{W}^\lambda) - 2\partial^\nu(\mathbf{W}^0\times\mathbf{W}^\lambda)]\cdot\mathbf{W}_{\nu\lambda}.
\end{aligned}$$

The final term here can be expressed in the form

$$\frac{1}{c}\mathbf{W}^{0\lambda}\cdot\left(\frac{g}{4\pi\hbar}\mathbf{W}^\nu\times\mathbf{W}_{\nu\lambda}\right) + \Delta = \frac{1}{c}\mathbf{W}^{0\lambda}\cdot(\mathbf{J}^W_\lambda - \mathbf{j}^W_\lambda) + \Delta.$$

The effect of this term is to remove the self-interaction current density $\mathbf{J}^W_\lambda - \mathbf{j}^W_\lambda$ from the dissipative term in the local conservation law. The residual invariant Δ can be shown to vanish identically. Using the identity

$$(\partial^\nu \mathbf{W}^\lambda) \times \mathbf{W}_{\nu\lambda} = \frac{1}{2}(\partial^\nu \mathbf{W}^\lambda - \partial^\lambda \mathbf{W}^\nu) \times \mathbf{W}_{\nu\lambda} \equiv \frac{g}{2\hbar c}(\mathbf{W}^\nu \times \mathbf{W}^\lambda) \times \mathbf{W}_{\nu\lambda},$$

the residual invariant is

$$\begin{aligned}\Delta &= \frac{g}{8\pi\hbar c}[2\partial^\nu(\mathbf{W}^0 \times \mathbf{W}^\lambda) - \partial^0(\mathbf{W}^\nu \times \mathbf{W}^\lambda) - 2\mathbf{W}^{0\lambda} \times \mathbf{W}^\nu] \cdot \mathbf{W}_{\nu\lambda} \\ &= \frac{g^2}{8\pi\hbar^2 c^2}\mathbf{W}^0 \cdot [2\mathbf{W}^\lambda \times (\mathbf{W}^\nu \times \mathbf{W}_{\nu\lambda}) + (\mathbf{W}^\nu \times \mathbf{W}^\lambda) \times \mathbf{W}_{\nu\lambda}] \\ &= \frac{g^2}{8\pi\hbar^2 c^2}\mathbf{W}^0 \cdot [\mathbf{W}^\nu(\mathbf{W}^\lambda \cdot \mathbf{W}_{\nu\lambda}) + \mathbf{W}^\lambda(\mathbf{W}^\nu \cdot \mathbf{W}_{\nu\lambda})] \equiv 0.\end{aligned}$$

The implied local conservation law for the interacting $SU(2)$ field is

$$\partial^\nu \Theta^0_\nu = -\frac{1}{c}\mathbf{W}^{0\lambda} \cdot \mathbf{j}^W_\lambda.$$

Further analysis is required to conclude that the local mass density must necessarily vanish.

10.5 The Standard Model

Because field quantization falls outside the scope of the present text, the discussion here has been limited to properties of classical fields that follow from Lorentz and general nonabelian gauge invariance of the Lagrangian densities. Treating the interacting fermion field as a classical field allows derivation of symmetry properties and of conservation laws, but is necessarily restricted to a theory of an isolated single particle. When this is extended by field quantization, so that the field amplitude ψ becomes a sum of fermion annihilation operators, the theory becomes applicable to the real world of many fermions and of physical antiparticles, while many qualitative implications of classical gauge field theory remain valid.

Quantized gauge fields play a central role in the widely accepted Standard Model of elementary particles and their interactions. Cottingham and Greenwood [73] provide an "elementary" introduction to the phenomological basis of this model and to the relevant theory, together with references to the technical literature. The postulate of generalized gauge invariance, found to imply renormalizability of the quantized theory, has powerful implications that are evident even for the classical fields. Fermion interactions have a strictly prescribed form, which determines the existence of gauge fields with specified internal symmetry, and determines corresponding conserved current densities. The Standard Model considers three such gauge fields, with fermion gauge transformations given by symmetry groups $U(1)$ (electromagnetic field), $SU(2)$ (weak isospin), and $SU(3)$ (strong color). There are three generations of fermion field. The generations (electron, muon, tauon) appear to differ only by the magnitude of the renormalized mass. The $U(1)$ field interacts with all fermions of nonvanishing electric charge; the $SU(2)$ field with all fermions (since isospin is quantized by half-integers for fermions); and the $SU(3)$ field (color gluons) interacts only with quark fermions, which carry a color charge. The combination $SU(2) \otimes (U(1) \oplus SU(3))$ requires eight elementary fermions in each generation.

The underlying symmetries of the fermion and gauge fields are broken dynamically in renormalized quantum field theory. In particular, helicity is broken by the Maxwell field, so that only neutrinos retain well-defined chirality. This implies a canonical transformation of the fermion vacuum state into a quasiparticle representation appropriate to the observed particles. As pointed out above, the Dirac equation is characteristic of such a representation, mixing bare massless fermions with different helicities. The differing physical masses of members of an isospin doublet, for example electron and neutrino, imply broken gauge symmetry. The $SU(2)$ gauge field Lagrangian of renormalized electroweak theory is decomposed into separate interaction terms for left- and right-handed components of the fermion field, since a massless neutrino has only left-handed helicity.

10.5.1 Electroweak theory (EWT)

The minimal internally consistent theory of elementary particles is a gauge theory of the elementary fermions of the first observed generation [340]: electron (e), electron-neutrino (ν_e), down quark (d), in three colors, and up quark (u), also in three colors. Strong interactions do not have to be considered if stable nuclei are treated as a passive background. If transitions between lepton generations are neglected, the theory can be applied independently to the lepton and quark doublets of each generation, interacting with the electromagnetic and weak gauge fields. A universal EWT field $e_0 U_\mu = \frac{1}{2}(g' y B_\mu + g \boldsymbol{\tau} \cdot \mathbf{W}_\mu)$ is coupled to fermion current densities, characterized by weak hypercharge y and weak isospin $\frac{1}{2}\boldsymbol{\tau}$, through the covariant derivative

$$D_\mu = \partial_\mu + i(e_0/\hbar c)U_\mu.$$

Here e_0 is the magnitude of the electronic charge. The equation of motion for the massless fermion field is

$$i\hbar c \gamma^\mu D_\mu \psi = 0.$$

Classical gauge fields, determined by their equations of motion, are expressed as Green function propagators acting on source current densities. The renormalized theory postulates a spontaneously broken symmetry as part of a dynamical mechanism that accounts for the large observed mass of the $SU(2)$ gauge field quanta. This mechanism implies a canonical transformation of the renormalized vacuum state that mixes the bare neutral fields W^3_μ and B_μ. Defining a Weinberg angle θ_W such that $g' \cos\theta_W = g \sin\theta_W = e_0$, the correct fermion electric charge $q_0 = \frac{1}{2}(y + \tau_3)e_0$ follows from an orthogonal transformation of the neutral fields

$$A_\mu = B_\mu \cos\theta_W \ + \ W^3_\mu \sin\theta_W,$$
$$Z^0_\mu = -B_\mu \sin\theta_W \ + \ W^3_\mu \cos\theta_W.$$

Defining $W^\pm_\mu = \frac{1}{\sqrt{2}}(W^1_\mu \pm W^2_\mu)$ and $\tau_\pm = \frac{1}{2}(\tau_1 \pm i\tau_2)$,

$$U_\mu = \frac{1}{2}(y + \tau_3)A_\mu + \frac{1}{2}(-y \tan\theta_W + \tau_3 \cot\theta_W)Z^0_\mu + \frac{1}{\sqrt{2}\sin\theta_W}(\tau_+ W^-_\mu + \tau_- W^+_\mu).$$

The parameter y is -1 for leptons and $1/3$ for quarks, while $\tau_3 = \mp 1$.

10.5.2 Quantum chromodynamics (QCD)

The $SU(3)$ group is the group of all 3×3 complex matrices with unit determinant. The condition $\det U = 1$ reduces the number of independent matrices to eight, so that there are eight independent gauge fields. These matrices act on quark fermion fields, which carry a color index with three values. The $SU(3)$ gluon fields correspond to matrices with two such indices, and are characterized either by a transition that changes color (six matrices) or by repeated color indices, in two independent matrices that have only diagonal elements. The Lagrangian density for this nonabelian gauge group implies that each gluon field contributes to the color current density. There is no empirical evidence for gluon field mass, in contrast to the observed large mass of the weak isospin $SU(2)$ fields.

Classical gauge fields determined by the equations [illegible] as Green function propagators [illegible] The [illegible] ized theory postulates spontaneous [illegible] symmetry [illegible] mechanism [illegible] field quanta [illegible] vacuum state [illegible]

[illegible]

[illegible]

[illegible]

References and bibliography

[1] Aashamar, K., Luke, T.M. and Talman, J.D. (1978). Optimized central potentials for atomic ground-state wavefunctions, *At. Data Nucl. Data Tables* **22**, 443–472.

[2] Abdolsalami, M. and Morrison, M.A. (1987). Calculating vibrational-excitation cross sections off the energy shell: A first-order adiabatic theory, *Phys. Rev. A* **36**, 5474–5477.

[3] Adhikari, S.K. (1998). *Variational Principles and the Numerical Solution of Scattering Problems* (Wiley, New York).

[4] Aissing, G. and Monkhorst, H.J. (1993). On the removal of the exchange singularity in extended systems, *Int. J. Quantum Chem.* **S27**, 81–89.

[5] Akhiezer, N.I. (1962). *The Calculus of Variations* (Blaisdell, New York).

[6] Amaldi, U. *et al.* (1987). Comprehensive analysis of data pertaining to the weak neutral current and the intermediate vector meson masses, *Phys. Rev. D* **36**, 1385–1407.

[7] Andersen, O.K. (1971). Comments on the KKR wavefunctions; extension of the spherical wave expansion beyond the muffin tins. In *Computational Methods in Band Theory*, eds. P.M. Marcus, J.F. Janak, A.R. Williams (Plenum, New York), 178–182.

[8] Andersen, O.K. (1973). Simple approach to the band-structure problem, *Solid State Commun.* **13**, 133–136.

[9] Andersen, O.K. (1975). Linear methods in band theory, *Phys. Rev. B* **12**, 3060–3083.

[10] Andersen, O.K. and Jepsen, O. (1977). Advances in the theory of one-electron states, *Physica B* **91**, 317–328.

[11] Andersen, O.K., Jepsen, O. and Glötzel, D. (1985). Canonical description of the band structures of metals. In *Highlights of Condensed-Matter Theory, Soc. Ital. Fis. Corso* **89**, 59–176.

[12] Andersen, O.K. and Kasowski, R.V. (1971). Electronic states as linear combinations of muffin-tin orbitals, *Phys. Rev. B* **4**, 1064–1069.

[13] Andersen, O.K. and Woolley, R.G. (1973). Muffin-tin orbitals and molecular calculations: General formalism, *Mol. Phys.* **26**, 905–927.

[14] Baluja, K., Burke, P.G. and Morgan, L.A. (1982). *R*-matrix propagation program for solving coupled second-order differential equations, *Comput. Phys. Commun.* **27**, 299–307.

[15] Bardsley, J.N. (1968). The theory of dissociative recombination, *J. Phys. B* **1**, 365–380.

[16] Bardsley, J.N., Herzenberg, A. and Mandl, F. (1966). Electron resonances of the H_2^- ion, *Proc. Phys. Soc. (London)* **89**, 305–320.
[17] Bardsley, J.N., Herzenberg, A. and Mandl, F. (1966). Vibrational excitation and dissociative attachment in the scattering of electrons by hydrogen molecules, *Proc. Phys. Soc. (London)* **89**, 321–340.
[18] Bardsley, J.N. and Mandl, F. (1968). Resonant scattering of electrons by molecules, *Rep. Prog. Phys.* **31**, 471–531.
[19] Bardsley, J.N. and Wadehra, J.M. (1979). Dissociative attachment and vibrational excitation in low-energy collisions of electrons with H_2 and D_2, *Phys. Rev. A* **20**, 1398–1405.
[20] Berman, M. and Domcke, W. (1984). Projection-operator calculations for shape resonances: A new method based on the many-body optical approach, *Phys. Rev. A* **29**, 2485–2496.
[21] Berman, M., Estrada, H., Cederbaum, L.S. and Domcke, W. (1983). Nuclear dynamics in resonant electron–molecule scattering beyond the local approximation: The 2.3-eV shape resonance in N_2, *Phys. Rev. A* **28**, 1363–1381.
[22] Berman, M., Mündel, C. and Domcke, W. (1985). Projection-operator calculations for molecular shape resonances: The $^2\Sigma_u^+$ resonance in electron-hydrogen scattering, *Phys. Rev. A* **31**, 641–651.
[23] Bingel, W.A. (1967). A physical interpretation of the cusp conditions for molecular wave functions, *Theor. Chim. Acta* **8**, 54–61.
[24] Birtwistle, D.T. and Herzenberg, A. (1971). Vibrational excitation of N_2 by resonance scattering of electrons, *J. Phys. B* **4**, 53–70.
[25] Bjorken, J.D. and Drell, S.C. (1965). *Relativistic Quantum Fields* (McGraw-Hill, New York).
[26] Blanchard, P. and Brüning, E. (1992). *Variational Methods in Mathematical Physics: A Unified Approach* (Springer-Verlag, Berlin).
[27] Blatt, J.M. and Biedenharn, L.C. (1952). The angular distribution of scattering and reaction cross sections, *Rev. Mod. Phys.* **24**, 258–272.
[28] Blatt, J.M. and Weisskopf, V.F. (1979). *Theoretical Nuclear Physics* (Springer-Verlag, New York).
[29] Bleecker, D. (1985). *Gauge Theory and Variational Principles* (Benjamin, New York).
[30] Bloch, C. (1957). A unified formulation of the theory of nuclear reactions, *Nucl. Phys.* **4**, 503–528.
[31] Born, M. and Oppenheimer, J.R. (1927). *Ann. Phys.* **84**, 457.
[32] Brandsden, B.H. (1970). *Atomic Collision Theory* (Plenum, New York).
[33] Breit, G. (1959). *Handbuch der Physik* **41**, No.1 (Springer, Berlin).
[34] Breit, G. and Wigner, E.P. (1936). Capture of slow neutrons, *Phys. Rev.* **49**, 519–531.
[35] Brenig, W. (1957). Zweiteilchennäherungen des Mehrkörperproblems I, *Nucl. Phys.* **4**, 363–374.
[36] Brescansin, L.M., Leite, J.R. and Ferreira, L.G. (1979). A study of the ground states and ionization energies of H_2, C_2, N_2, F_2, and CO molecules by the variational cellular method, *J. Chem. Phys.* **71**, 4923–4930.
[37] Brillouin, L. (1934). *Actual. Sci. Ind.* **159**, 28.
[38] Brown, L.M. (1958). Two-component fermion theory, *Phys. Rev.* **111**, 957–964.
[39] Brown, R.G. and Ciftan, M. (1983). Generalized non-muffin-tin band theory, *Phys. Rev. B* **27**, 4564–4579.
[40] Brown, R.G. and Ciftan, M. (1983). *Phys. Rev. B* **28**, 5992–6007.

[41] Broyden, C.G. (1967). Quasi-Newton methods and their application to function minimisation, *Math. Comput.* **21**, 368–381.
[42] Broyden, C.G. (1970). The convergence of single-rank quasi-Newton methods, *Math. Comput.* **24**, 365–382.
[43] Brueckner, K.A. and Wada, W. (1956). Nuclear saturation and two-body forces: Self-consistent solutions and the effects of the exclusion principle, *Phys. Rev.* **103**, 1008–1016.
[44] Burke, P.G. (1973). The *R*-matrix method in atomic physics, *Comput. Phys. Commun.* **6**, 288–302.
[45] Burke, P.G. (1977). *Potential Scattering in Atomic Physics* (Plenum, New York).
[46] Burke, P.G. and Robb, W.D. (1972). Elastic scattering of electrons by hydrogen and helium atoms, *J. Phys. B* **5**, 44–54.
[47] Burke, P.G. and Robb, W.D. (1975). The *R*-matrix theory of atomic processes, *Adv. At. Mol. Phys.* **11**, 143–214.
[48] Burke, P.G. and Seaton, M.J. (1971). Numerical solutions of the integro-differential equations of electron–atom collision theory, *Methods Comput. Phys.* **10**, 1–80.
[49] Burke, P.G. and West, J.B., eds. (1987). *Electron–Molecule Scattering and Photoionization* (Plenum, New York).
[50] Burke, P.G, Hibbert, A. and Robb, W.D. (1971). Electron scattering by complex atoms, *J. Phys. B* **4**, 153–161.
[51] Butler, W.H. and Nesbet, R.K. (1990). Validity, accuracy, and efficiency of multiple-scattering theory for space-filling scatterers, *Phys. Rev. B* **42**, 1518–1525.
[52] Butler, W.H. and Zhang, X.-G. (1991). Accuracy and convergence properties of multiple-scattering theory in three dimensions, *Phys. Rev. B* **44**, 969–983.
[53] Butler, W.H., Zhang, X.-G. and Gonis, A. (1992). The Green function cellular method and its relation to multiple scattering theory, *Mat. Res. Symp. Proc.* **253**, 205–210.
[54] Buttle, P.J.A. (1967). Solution of coupled equations by *R*-matrix techniques, *Phys. Rev.* **160**, 719–729.
[55] Cairo, L. (1965). *Variational Techniques in Electromagnetism*, tr. G.D. Sims (Gordon and Breach, New York).
[56] Callaway, J. and March, N.H. (1985). Density functional methods: theory and applications, *Solid State Phys.* **38**, 135–221.
[57] Callen, H.B. (1963). *Thermodynamics* (Wiley, New York).
[58] Carathéodory, C. (1967). *Calculus of Variations and Partial Differential Equations of the First Order*, Part II, tr. R.B. Dean and J.J. Brandstatter (Holden-Day, San Francisco).
[59] Cederbaum, L.S. and Domcke, W. (1981). Local against nonlocal complex potential in resonant electron–molecule scattering, *J. Phys. B* **14**, 4665–4689.
[60] Ceperley, D.M. and Alder, B.J. (1980). Ground state of the electron gas by a stochastic method, *Phys. Rev. Lett.* **45**, 566–569.
[61] Chandra, N. and Temkin, A. (1976). Hybrid theory and calculation of $e-N_2$ scattering, *Phys. Rev. A* **13**, 188–215.
[62] Chang, E.S. and Fano, U. (1972). Theory of electron–molecule collisions by frame transformation, *Phys. Rev. A* **6**, 173–185.
[63] Chang, E.S. and Temkin, A. (1969). Rotational excitation of diatomic molecules by electron impact, *Phys. Rev. Lett.* **23**, 399–403.
[64] Chang, E.S. and Temkin, A. (1970). Rotational excitation of diatomic molecular systems. II. H_2^+, *J. Phys. Soc. Japan* **29**, 172–179.

[65] Chase, D.M. (1956). Adiabatic approximation for scattering processes, *Phys. Rev.* **104**, 838–842.

[66] Chatwin, R.A. and Purcell, J.E. (1971). Approximate solution of a Sturm–Liouville system using nonorthogonal expansions: Application to α–α nuclear scattering, *J. Math. Phys.* **12**, 2024–2030.

[67] Chen, J.C.Y. (1966). Dissociative attachment in rearrangement electron collision with molecules, *Phys. Rev.* **148**, 66–73.

[68] Clementi, E. and Roetti, C. (1974). Roothaan–Hartree–Fock atomic wavefunctions, *At. Data Nucl. Data Tables* **14**, 177–478.

[69] Colle, R. and Nesbet, R.K. (2001). Optimized effective potential in finite-basis-set treatment, *J. Phys. B* **34**, 2475–2480.

[70] Colle, R. and Salvetti, O. (1975). Approximate calculation of the correlation energy for the closed shells, *Theoret. Chim. Acta* **37**, 329–334.

[71] Collins, J.C. (1984). *Renormalization* (Cambridge University Press, Cambridge).

[72] Condon, E.U. and Shortley, G.H. (1935). *The Theory of Atomic Spectra* (Cambridge University Press, Cambridge).

[73] Cottingham, W.N. and Greenwood, D.A. (1998). *An Introduction to the Standard Model of Particle Physics* (Cambridge University Press, New York).

[74] Courant, R. and Hilbert, D. (1953). *Methods of Mathematical Physics* (Interscience, New York).

[75] Csaszar, P. and Pulay, P. (1984). Geometry optimization by direct inversion in the iterative subspace, *J. Mol. Struct.* **114**, 31–34.

[76] Damburg, R.J. and Karule, E. (1967). A modification of the close-coupling approximation for e – H scattering allowing for the long-range interaction, *Proc. Phys. Soc. (London)* **90**, 637–640.

[77] Dennis, J.E. and Schnabel, R.B. (1983). *Numerical Methods for Unconstrained Optimization and Nonlinear Equations* (Prentice-Hall, Englewood Cliffs, New Jersey).

[78] Dieudonné, J. (1981). *History of Functional Analysis* (North-Holland, Amsterdam).

[79] Dirac, P.A.M. (1930). Note on exchange phenomena in the Thomas atom, *Proc. Camb. Phil. Soc.* **26**, 376–385.

[80] Dirac, P.A.M. (1933). *Physikalische Z. der Sowjetunion* **3**, 64.

[81] Dirac, P.A.M. (1958). *Quantum Mechanics*, 4th edition (Oxford University Press, London).

[82] Domcke, W. (1991). Theory of resonance and threshold effects in electron–molecule collisions: the projection-operator approach, *Phys. Rep.* **208**, 97–188.

[83] Domcke, W. and Cederbaum, L.S. (1977). Theory of the vibrational structure of resonances in electron–atom scattering, *Phys. Rev. A* **16**, 1465–1482.

[84] Domcke, W. and Cederbaum, L.S. (1981). On the interpretation of low-energy electron-HCl scattering phenomena, *J. Phys. B* **14**, 149–173.

[85] Domcke, W., Cederbaum, L.S. and Kaspar, F. (1979). Threshold phenomena in electron–molecule scattering: a nonadiabatic theory, *J. Phys. B* **12**, L359–L364.

[86] Domcke, W. and Mündel, C. (1985). Calculation of cross sections for vibrational excitation and dissociative attachment in HCl and DCl beyond the local complex potential approximation, *J. Phys. B* **18**, 4491–4509.

[87] Domps, A., Reinhard, P.-G. and Suraud, E. (1998). Time-dependent Thomas–Fermi approach to electron dynamics in metal clusters, *Phys. Rev. Lett.* **80**, 5520–5523.

[88] Dovesi, R., Saunders, V.R., Roetti, C., Causà, M., Harrison, N.M., Orlando, R. and Aprà, E. (1996). *CRYSTAL95 User's Manual* (University of Torino, Torino, Italy).

[89] Dreizler, R.M. and de Providencia, J. (1985). *Density Functional Methods in Physics* (Plenum, New York).
[90] Dreizler, R.M. and Gross, E.K.U. (1990). *Density Functional Theory* (Springer-Verlag, Berlin).
[91] Dubé, L. and Herzenberg, A. (1979). Absolute cross sections from the "boomerang model" for resonant electron–molecule scattering, *Phys. Rev. A* **20**, 194–213.
[92] Durrant, A.V. (1980). A derivation of optical field quantization from absorber theory, *Proc. Roy. Soc. (London)* **A370**, 41–59.
[93] Edelen, D.G.B. (1969). *Nonlocal Variations and Local Invariance of Fields* (American Elsevier, New York).
[94] Ehrenreich, H. and Cohen, M.H. (1959). Self-consistent field approach to the many-electron problem, *Phys. Rev.* **115**, 786–790.
[95] Ehrhardt, H. and Willmann, K. (1967). Angular dependence of low-energy resonance scattering of electrons by N_2, *Z. Phys.* **204**, 462–473.
[96] Elsgolc, L.E. (1962). *Calculus of Variations* (Pergamon Press, London).
[97] Engel, E. and Vosko, S.H. (1993). Accurate optimized-potential-model solutions for spherical spin-polarized atoms: Evidence for limitations of the exchange-only local spin-density and generalized-gradient expansions, *Phys. Rev. A* **47**, 2800–2811.
[98] Englert, F. and Brout, R. (1964). Broken symmetry and the mass of gauge vector mesons, *Phys. Rev. Lett.* **13**, 321–323.
[99] Englisch, H. and Englisch, R. (1984). Exact density functionals for ground-state energies. I General results, *Phys. Stat. Sol. B* **123**, 711–721.
[100] Englisch, H. and Englisch, R. (1984). Exact density functionals for ground-state energies. II Details and remarks, *Phys. Stat. Sol. B* **124**, 373–379.
[101] Epstein, S.T. (1974). *The Variation Method in Quantum Chemistry* (Academic Press, New York).
[102] Eschrig, H. (1996). *The Fundamentals of Density Functional Theory* (Teubner, Stuttgart).
[103] Fabrikant, I.I. (1990). Resonance processes in e-HCl collisions: comparison of the *R*-matrix and the nonlocal-complex-potential methods, *Comments At. Mol. Phys.* **24**, 37–52.
[104] Faddeev, L.D. and Slavnov, A.A. (1980). *Gauge Fields: Introduction to Quantum Theory* (Benjamin/Cummings, Reading, Massachusetts).
[105] Faisal, F.H.M. and Temkin, A. (1972). Application of the adiabatic-nuclei theory to vibrational excitation, *Phys. Rev. Lett.* **28**, 203–206.
[106] Fano, U. and Lee, C.M. (1973). Variational calculation of *R*-matrices. Application to Ar photoabsorption, *Phys. Rev. Lett.* **31**, 1573–1576.
[107] Faulkner, J.S. (1986). Non-muffin-tin potentials in multiple scattering theory, *Phys. Rev. B* **34**, 5931–5934.
[108] Fermi, E. (1927). Un metodo statistice per la determinazione di alcune proprieta dell'atomo, *Rend. Accad. Lincei* **6**, 602–607.
[109] Fermi, E. (1928). A statistical method for the determination of some atomic properties and the application of this method to the theory of the periodic system of elements, *Z. Phys.* **48**, 73–79.
[110] Ferraz, A.C., Chagas, M.I.T., Takahashi, E.K. and Leite, J.R. (1984). Variational cellular model of the energy bands of diamond and silicon, *Phys. Rev. B* **29**, 7003–7006.
[111] Ferraz, A.C., Takahashi, E.K. and Leite, J.R. (1982). Application of the variational cellular method to periodic structures. Energy band of sodium, *Phys. Rev. B* **26**, 690–700.

[112] Ferreira, L.G. and Leite, J.R. (1978). Variational cellular model of the molecular and crystal electronic structure, *Phys. Rev. Lett.* **40**, 49–52.
[113] Ferreira, L.G. and Leite, J.R. (1978). General formulation of the variational cellular method for molecules and crystals, *Phys. Rev. A* **18**, 335–343.
[114] Ferreira, L.G. and Leite, J.R. (1979). Self-consistent calculation of the electronic structure of N_2 and CO by the variational cellular method, *Phys. Rev. A* **20**, 689–699.
[115] Feshbach, H. (1958). Unified theory of nuclear reactions, *Ann. Phys. (New York)* **5**, 357–390.
[116] Feshbach, H. (1962). A unified theory of nuclear reactions, *Ann. Phys. (New York)* **19**, 287–313.
[117] Feynman, R.P. (1939). Forces in molecules, *Phys. Rev.* **56**, 340–343.
[118] Feynman, R.P. (1948). A relativistic cut-off for classical electrodynamics, *Phys. Rev.* **74**, 939–946.
[119] Feynman, R.P. (1948). Space-time approach to nonrelativistic quantum mechanics, *Rev. Mod. Phys.* **20**, 367–387.
[120] Feynman, R.P. (1949). Space-time approach to quantum electrodynamics, *Phys. Rev.* **76**, 769–789.
[121] Feynman, R.P. (1961). *Quantum Electrodynamics* (Benjamin, New York).
[122] Feynman, R.P. and Gell-Mann, M. (1958). Theory of the Fermi interaction, *Phys. Rev.* **109**, 193–198.
[123] Feynman, R.P. and A.R. Hibbs (1965). *Quantum Mechanics and Path Integrals* (McGraw-Hill, New York).
[124] Fletcher, R. (1970). A new approach to variable metric algorithms, *Comput. J.* **13**, 317–322.
[125] Fletcher, R. (1987). *Practical Methods of Optimization*, 2nd edition (Wiley, New York).
[126] Fliflet, A.W. and McKoy, V. (1978). Discrete basis set method for electron–molecule continuum wave functions, *Phys. Rev. A* **18**, 2107–2114.
[127] Fock, V. (1930). *Z. Phys.* **61**, 126.
[128] Fock, V. (1957). *The Theory of Space, Time and Gravitation*, tr. N. Kemmer (Pergamon Press, London).
[129] Fried, H. (1972). *Functional Methods and Models in Quantum Field Theory* (MIT Press, Cambridge, Massachusetts).
[130] Froese Fischer, C. (1977). *The Hartree–Fock Method for Atoms* (Wiley, New York).
[131] Garrod, G. and Percus, J.K. (1964). Reduction of the N-particle variational problem, *J. Math. Phys.* **5**, 1756–1776.
[132] Gerjuoy, E., Rau, A.R.P. and Spruch, L. (1983). A unified formulation of the construction of variational principles, *Rev. Mod. Phys.* **55**, 725–774.
[133] Gianturco, F.A. and Stoecklin, T. (1996). The elastic scattering of electrons from CO_2 molecules: I. Close coupling calculations of integral and differential cross sections, *J. Phys. B* **29**, 3933–3954.
[134] Gianturco, F.A., Jain, A. and Pantano, L. (1987). Electron-methane scattering via a parameter-free model interaction, *J. Phys. B* **20**, 571–586.
[135] Gianturco, F.A., Thompson, D.G. and Jain, A. (1995). Electron scattering from polyatomic molecules using a single-center expansion formulation, in *Computational Methods for Electron–Molecule Collisions*, eds. W.M. Huo and F.A. Gianturco (Plenum, New York), pp. 75–118.
[136] Gilbert, T.L. (1975). Hohenberg–Kohn theorem for nonlocal external potentials, *Phys. Rev. B* **12**, 2111–2120.

[137] Gill, P.E., Murray, W. and Wright, M.H. (1981). *Practical Optimization* (Academic Press, New York).

[138] Gillan, C.J., Nagy, O., Burke, P.G., Morgan, L.A. and Noble, C.J. (1987). Electron scattering by nitrogen molecules, *J. Phys. B* **20**, 4585–4603.

[139] Gillan, C.J., Tennyson, J. and Burke, P.G. (1995). The UK molecular *R*-matrix scattering package: a computational perspective, in *Computational Methods for Electron–Molecule Collisions*, eds. W.M. Huo and F.A. Gianturco (Plenum, New York), pp. 239–254.

[140] Glashaw, S.L. (1961). Partial symmetries of weak interactions, *Nucl. Phys.* **22**, 579–588.

[141] Glimm, A. and Jaffe, A. (1981). *Quantum Mechanics - A Functional Integral Point of View* (Springer-Verlag, Berlin).

[142] Glowinski, R. (1983). *Numerical Methods for Nonlinear Variational Problems* (Springer-Verlag, Berlin).

[143] Golden, D.E. (1966). Low-energy resonances in $e^- - N_2$ total scattering cross sections: the temporary formation of N_2^-, *Phys. Rev. Lett.* **17**, 847–851.

[144] Golden, D.E., Lane, N.F., Temkin, A. and Gerjuoy, E. (1971). Low energy electron–molecule scattering experiments and the theory of rotational excitation, *Rev. Mod. Phys.* **43**, 642–678.

[145] Goldfarb, D. (1970). A family of variable-metric methods derived by variational means, *Math. Comput.* **24**, 23–26.

[146] Goldstein, H. (1983). *Classical Mechanics*, 2nd edition (Wiley, New York).

[147] Goldstine, H.H. (1980). *A History of the Calculus of Variations from the 17th Through the 19th Century* (Springer-Verlag, Berlin).

[148] Gonis, A. and Butler, W.H. (2000). *Multiple Scattering in Solids* (Springer, New York).

[149] Gonze, X. and Scheffler, M. (1999). Exchange and correlation kernels at the resonance frequency: implications for excitation energies in density-functional theory, *Phys. Rev. Lett.* **82**, 4416–4419.

[150] Gordon, R.G. and Kim, Y.S. (1972). Theory for the forces between closed-shell atoms and molecules, *J. Chem. Phys.* **56**, 3122–3133.

[151] Görling, A. (1999). New KS method for molecules based on an exchange charge density generating the exact KS exchange potential, *Phys. Rev. Lett.* **83**, 5459–5462.

[152] Görling, A. and Ernzerhof, M. (1995). Energy differences between Kohn–Sham and Hartree–Fock wave functions yielding the same electron density, *Phys. Rev. A* **51**, 4501–4513.

[153] Grimm-Bosbach, T., Thümmel, H.T., Nesbet, R.K. and Peyerimhoff, S.D. (1996). Calculation of cross sections for rovibrational excitation of N_2 by electron impact, *J. Phys. B* **29**, L105–L112.

[154] Gross, E.K.U. and Kohn, W. (1985). Local density-functional theory of frequency-dependent linear response, *Phys. Rev. Lett.* **55**, 2850–2852.

[155] Gross, E.K.U., Dobson, J.F. and Petersilka, M. (1996). Density functional theory of time-dependent phenomena, in *Density Functional Theory*, ed. R.F. Nalewajski, Series "Topics in Current Chemistry" (Springer, Berlin).

[156] Guralnik, G.S., Hagen, C.R. and Kibble, T.W.B. (1964). Global conservation laws and massless particles, *Phys. Rev. Lett.* **13**, 585–587.

[157] Gyorffy, B.L. (1972). Coherent-potential approximation for a nonoverlapping muffin-tin potential model of random substitutional alloys, *Phys. Rev. B* **5**, 2382–2384.

[158] Ham, F. and Segal, B. (1961). Energy bands in periodic lattices – Green's function method, *Phys. Rev.* **124**, 1786–1796.
[159] Harris, J. (1979). The role of occupation numbers in HKS theory, *Int. J. Quantum Chem.* **S13**, 189–192.
[160] Harris, J. (1984). Adiabatic-connection approach to Kohn–Sham theory, *Phys. Rev. A* **29**, 1648–1659.
[161] Harris, J. and Jones, R.O. (1974). The surface energy of a bounded electron gas, *J. Phys. F* **4**, 1170–1186.
[162] Hartree, D.H. (1928). The wave mechanics of an atom with a non-Coulomb central field. Part II. Some results and discussion, *Proc. Camb. Phil. Soc.* **24**, 111–132.
[163] Hartree, D.H. (1957). *The Calculation of Atomic Structures* (Wiley, New York).
[164] Hazi, A.U., Orel, A.E. and Rescigno, T.N. (1981). *Ab initio* study of dissociative attachment of low-energy electrons to F_2, *Phys. Rev. Lett.* **46**, 918–922.
[165] Head, J.D., Weiner, B. and Zerner, M.C. (1988). A survey of optimization procedures for stable structures and transition states, *Int. J. Quantum Chem.* **33**, 177–186.
[166] Head, J.D. and Zerner, M.C. (1988). Newton-based optimization methods for obtaining molecular conformation, *Adv. Quantum Chem.* **20**, 241–290.
[167] Herzenberg, A. (1968). Oscillatory energy dependence of resonant electron–molecule scattering, *J. Phys. B* **1**, 548–558.
[168] Hestenes, M.R. (1966). *Calculus of Variations and Optimal Control Theory* (Wiley, New York).
[169] Higgs, P.W. (1964). Broken symmetries and the masses of gauge bosons, *Phys. Rev. Lett.* **13**, 508–509.
[170] Higgs, P.W. (1966). Spontaneous symmetry breakdown without massless bosons, *Phys. Rev.* **145**, 1156–1163.
[171] Hill, E.L. (1951). Hamilton's principle and the conservation theorems of mathematical physics, *Rev. Mod. Phys.* **23**, 253–260.
[172] Höck, A. and Engel, E. (1998). Pseudopotentials from orbital-dependent exchange-correlation functionals, *Phys. Rev. A* **58**, 3578–3581.
[173] Hohenberg, P. and Kohn, W. (1964). Inhomogeneous electron gas, *Phys. Rev.* **136**, B864–B871.
[174] Hoyle, F.M. and Narlikar, J.V. (1974). *Action at a Distance in Physics and Cosmology* (Freeman, San Francisco).
[175] Hulthén, L. (1944). *Kgl. Fysiogr. Sällsk. Lund Förh.* **14**, No.21.
[176] Hulthén, L. (1948). *Arkiv. Mat. Astron. Fysik* **35A**, No.25.
[177] Huo, W.M. (1995). The Schwinger variational method, in *Computational Methods for Electron–Molecule Collisions*, eds. W.M. Huo and F.A. Gianturco (Plenum, New York), pp. 326–355.
[178] Huo, W.M. and Gianturco, F.A. (1995). *Computational Methods for Electron–Molecule Collisions* (Plenum, New York).
[179] Hurley, A.C. (1954). The electrostatic calculation of molecular energies II. Approximate wave functions and the electrostatic method, *Proc. Roy. Soc. A* **226**, 179–192.
[180] Hurley, A.C. (1976). *Electron Correlation in Small Molecules* (Academic Press, New York).
[181] Ivanov, S., Hirata, S. and Bartlett, R.J. (1999). Exact exchange treatment for molecules in finite-basis-set Kohn–Sham theory, *Phys. Rev. Lett.* **83**, 5455–5458.
[182] Jackson, J.D. (1975). *Classical Electrodynamics*, 2nd edition (Wiley, New York).

[183] Jackson, J.L. (1951). A variational appeoach to nuclear reactions, *Phys. Rev.* **83**, 301–304.
[184] Jacob, M. and Wick, G.C. (1959). On the general theory of collisions for particles with spin, *Ann. Phys. (New York)* **7**, 404–428.
[185] Janak, J.F. (1978). Proof that $\partial E/\partial n_i = \epsilon_i$ in density-functional theory, *Phys. Rev. B* **18**, 7165–7168.
[186] Jauch, J.M. and Rohrlich, F. (1985). *The Theory of Photons and Electrons* (Springer-Verlag, Berlin).
[187] Jeffreys, H. and Jeffreys, B.S. (1956). *Methods of Mathematical Physics*, 3rd edition (Cambridge University Press, New York).
[188] Joachain, C.J. (1975). *Quantum Collision Theory*, Vol.1 (North-Holland, Amsterdam).
[189] Jones, R.O. and Gunnarsson, O. (1989). The density functional formalism, its applications and prospects, *Rev. Mod. Phys.* **61**, 689–746.
[190] Itzykson, C. and Zuber, J.-B. (1980). *Quantum Field Theory* (McGraw-Hill, New York).
[191] Kamimura, M. (1977). A coupled channel variational method for microscopic study of reactions between complex nuclei, *Prog. Theor. Phys. Suppl.* **62**, 236–294.
[192] Kapur, P.L. and Peierls, R. (1938). The dispersion formula for nuclear reactions, *Proc. Roy. Soc. (London)* **A166**, 277–295.
[193] Kasowski, R.V. and Andersen, O.K. (1972). Muffin tin orbitals in open structures, *Solid State Commun.* **11**, 799–802.
[194] Kato, T. (1950). Variational methods in collision problems, *Phys. Rev.* **80**, 475.
[195] Kato, T. (1951). Upper and lower bounds of scattering phases, *Prog. Theor. Phys.* **6**, 394–407.
[196] Kato, T. (1957). On the eigenfunctions of many-particle systems in quantum mechanics, *Commun. Pure Appl. Math.* **10**, 151–177.
[197] Kaufmann, K., Baumeister, W. and Jungen, M. (1989). Universal Gaussian basis sets for an optimum representation of Rydberg and continuum wavefunctions, *J. Phys. B* **22**, 2223–2240.
[198] Kibble, T.W.B. (1967). Symmetry breaking in non-abelian gauge theories, *Phys. Rev.* **155**, 1554–1561.
[199] Klein, F. (1918). *Nachr. Ges. Wiss. Göttingen*, 171.
[200] Kline, M. (1972). *Mathematical Thought from Ancient to Modern Times*, 3 vols. (Oxford University Press, New York).
[201] Kohl, H. and Dreizler, R.M. (1986). Time-dependent density-functional theory: conceptual and practical aspects, *Phys. Rev. Lett.* **56**, 1993–1995.
[202] Kohn, W. (1948). Variational methods in nuclear collision problems, *Phys. Rev.* **74**, 1763–1772.
[203] Kohn, W. (1952). Variational methods for periodic lattices, *Phys. Rev.* **87**, 472–481.
[204] Kohn, W. and Rostoker, N. (1954). Solution of the Schrödinger equation in periodic lattices with an application to metallic lithium, *Phys. Rev.* **94**, 1111–1120.
[205] Kohn, W. and Sham, L.J. (1965). Self consistent equations including exchange and correlation effects, *Phys. Rev.* **140**, A1133–A1138.
[206] Koopmans, T. (1933). Über die Zuordnung von Wellenfunktionen und Eigenwerten zu den einzelnen Elektronen eines Atoms, *Physica* **1**,104–113.
[207] Korringa, J. (1947). On the calculation of the energy of a Bloch wave in a metal, *Physica* **13**, 392–400.
[208] Kuperschmidt, B.A. (1990). *The Variational Principles of Dynamics* (World Scientific, New York).

[209] Lai, C.H. (1981). *Gauge Theory of Electromagnetic and Weak Interactions*, ed. C.H. Lai (World Scientific, Singapore).

[210] Lanczos, C. (1966). *Variational Principles of Mechanics* (University of Toronto Press, Toronto).

[211] Landau, L.D. (1956). The theory of a Fermi liquid, *Zh. Eksp. Teor. Fiz.* **30**, 1058–1064. [*Sov. Phys. JETP* **3**, 920–925 (1956)]

[212] Landau, L.D. (1957). Oscillations in a Fermi liquid, *Zh. Eksp. Teor. Fiz.* **32**, 59–66. [*Sov. Phys. JETP* **5**, 101–108 (1957)]

[213] Landau, L.D. and Lifshitz, E.M. (1958). *Quantum Mechanics, Non-Relativistic Theory*, translated by J.B. Sykes and J.S. Bell (Pergamon, London).

[214] Lane, A.M. and Thomas, R.G. (1958). R-matrix theory of nuclear reactions, *Rev. Mod. Phys.* **30**, 257–353.

[215] Lane, N.F. (1980). The theory of electron–molecule collisions, *Rev. Mod. Phys.* **52**, 29–119.

[216] Langreth, D.C. and Mehl, M.J. (1981). Easily implementable nonlocal exchange-correlation energy functional, *Phys. Rev. Lett.* **47**, 446–450.

[217] Leader, E. and Predazzi, E. (1996). *An Introduction to Gauge Theories and Modern Particle Physics*, Vols. 1 and 2 (Cambridge University Press, New York).

[218] Lee, B.W. and Zinn-Justin, J. (1972). Spontaneously broken gauge symmetries. I. Preliminaries, *Phys. Rev. D* **5**, 3121–3137.

[219] Lee, C., Yang, W. and Parr, R.G. (1988). Development of the Colle-Salvetti correlation energy formula into a functional of the electron density, *Phys. Rev. B* **37**, 785–789.

[220] Lee, C.M. (1974). Spectroscopy and collision theory. III. Atomic eigenchannel calculation by a Hartree–Fock-Roothaan method, *Phys. Rev. A* **10**, 584–600.

[221] Leibbrandt, G. (1987). Introduction to noncovariant gauges, *Rev. Mod. Phys.* **59**, 1067–1119.

[222] Levy, M. (1979). Universal variational functionals of electron densities, first-order density matrices, and natural spin-orbitals and solution of the v-representability problem, *Proc. Natl. Acad. Sci.* **76**, 6062–6065.

[223] Lieb, E.H. (1983). Density functionals for Coulomb systems, *Int. J. Quantum Chem.* **24**, 243–277.

[224] Lieb, E.H. and Simon, B. (1977). The Thomas–Fermi theory of atoms, molecules and solids, *Adv. Math.* **23**, 22–216.

[225] Light, J.C. and Walker, R.B. (1976). An R-matrix approach to the solution of coupled equations for atom–molecule scattering, *J. Chem. Phys.* **65**, 4272–4282.

[226] Lima, M.A.P., Leite, J.R. and Fazio, A. (1971). Theoretical study of the F_2 molecule using the variational cellular method, *J. Phys. B* **14**, L533–L535.

[227] Lino, A.T., Leite, J.R., Ferraz, A.C. and Takahashi, E.K. (1987). Self-consistent formulation of the variational cellular method applied to periodic structures – results for sodium and silicon, *J. Phys. Chem. Solids* **48**, 911–919.

[228] Lippmann, B.A. and Schwinger, J. (1950). Variational principles for scattering processes. I, *Phys. Rev.* **79**, 469–480.

[229] Loucks, T.L. (1967). *Augmented Plane Wave Method* (Benjamin, New York), pp. 98–103.

[230] Lucchese, R.R., Takatsuka, K. and McKoy, V. (1986). Applications of the Schwinger variational principle to electron–molecule collisions and molecular photoionization, *Phys. Rep.* **131**, 147–221.

[231] March, N.H. (1957). The Thomas–Fermi approximation in quantum mechanics, *Adv. Phys.* **6**, 1–101.

[232] March, N.H. (1975). *Self-consistent Fields in Atoms* (Pergamon Press, Oxford).
[233] Mazevet, S., Morrison, M.A., Boydstun, O. and Nesbet, R.K. (1999). Inclusion of nonadiabatic effects in calculations on vibrational excitation of molecular hydrogen by low-energy electron impact, *Phys. Rev. A* **59**, 477–489.
[234] Mazevet, S., Morrison, M.A., Boydstun, O. and Nesbet, R.K. (1999). Adiabatic treatments of vibrational dynamics in low-energy electron–molecule scattering, *J. Phys. B* **32**, 1269–1294.
[235] Mazevet, S., Morrison, M.A., Morgan, L.A. and Nesbet, R.K. (2001). Virtual-state effects on elastic scattering and vibrational excitation of CO_2 by electron impact. *Phys. Rev. A* **64**, 040701.
[236] Mazevet, S., Morrison, M.A. and Nesbet, R.K. (1998). Application of the nonadiabatic phase matrix method to vibrational excitation near a short-lived resonance: the case of e-H_2 scattering, *J. Phys. B* **31**, 4437–4448.
[237] McCurdy, C.W., Rescigno, T.N. and Schneider, B.I. (1987). Interrelation between variational principles for scattering amplitudes and generalized R-matrix theory, *Phys. Rev. A* **36**, 2061–2066.
[238] McLachlan, A.D. and Ball, M.A. (1964). Time-dependent Hartree–Fock theory for molecules, *Rev. Mod. Phys.* **36**, 844–855.
[239] McWeeny, R. (1989). *Methods of Molecular Quantum Mechanics* (Academic Press, New York).
[240] Mercier, A. (1959). *Analytical and Canonical Formalism in Physics* (Interscience, New York).
[241] Mermin, N.D. (1965). Thermal properties of the inhomogeneous electron gas, *Phys. Rev.* **137**, A1441–A1443.
[242] Merzbacher, E. (1961). *Quantum Mechanics* (Wiley, New York).
[243] Messiah, A. (1961). *Quantum Mechanics*, Vol. I (North Holland, Amsterdam).
[244] Miller, W.H. and Jansen op de Haar, B.M.D.D. (1987). A new basis set for quantum scattering calculations, *J. Chem. Phys.* **86**, 6213–6220.
[245] Mito, Y. and Kamimura, M. (1976). The generator coordinate method for composite-particle scattering based on the Kohn-Hulthén variational principle, *Prog. Theor. Phys.* **56**, 583–598.
[246] Moiseiwitsch, B.L. (1966). *Variational Principles* (Interscience, New York).
[247] Morgan, J. van W. (1977). Integration of Poisson's equation for a complex system with arbitrary geometry, *J. Phys. C* **10**, 1181–1202.
[248] Morgan, L.A. (1984). A generalized R-matrix propagation program for solving coupled second-order differential equations, *Comput. Phys. Commun.* **31**, 419–422.
[249] Morgan, L.A. (1998). Virtual states and resonances in electron scattering by CO_2, *Phys. Rev. Lett.* **80**, 1873–1875.
[250] Morgan, L.A. and Burke, P.G. (1988). Low-energy electron scattering by HF, *J. Phys. B* **21**, 2091–2105.
[251] Morgan, L.A., Burke, P.G. and Gillan, C.J. (1990). Low-energy electron scattering by HCl, *J. Phys. B* **23**, 99–113.
[252] Morrison, M.A. (1986). A first-order nondegenerate adiabatic theory for calculating near-threshold cross sections for rovibrational excitation of molecules by electron impact, *J. Phys. B* **19**, L707–L715.
[253] Morrison, M.A. (1988). Near-threshold electron–molecule scattering, *Adv. At. Mol. Phys.* **24**, 51–156.
[254] Morrison, M.A., Abdolsalami, M. and Elza, B.K. (1991). Improved accuracy in adiabatic cross sections for low-energy rotational and vibrational excitation, *Phys. Rev. A* **43**, 3440–3459.

[255] Morse, P.M. and Feshbach, H. (1953). *Methods of Theoretical Physics*, Vols. I and II (McGraw-Hill, New York).
[256] Moruzzi, V.L., Janak, J.F. and Williams, A.R. (1978). *Calculated Electronic Properties of Metals* (Pergamon, New York).
[257] Mott, N.F. and Massey, H.S.W. (1965). *The Theory of Atomic Collisions* (Oxford University Press, New York).
[258] Mündel, C. and Domcke, W. (1984). Nuclear dynamics in resonant electron–molecule scattering beyond the local approximation: model calculations on dissociative attachment and vibrational excitation, *J. Phys. B* **17**, 3593–3616.
[259] Mündel, C., Berman, M. and Domcke, W. (1985). Nuclear dynamics in resonant electron–molecule scattering beyond the local approximation: Vibrational excitation and dissociative attachment in H_2 and D_2, *Phys. Rev. A* **32**, 181–193.
[260] Nagy, A. (1995). Coordinate scaling and adiabatic connection formula for ensembles of fractionally occupied excited states, *Int. J. Quantum Chem.* **56**, 225–228.
[261] Nesbet, R.K. (1955). Configuration interaction in orbital theories, *Proc. Roy. Soc. A* **230**, 312–321.
[262] Nesbet, R.K. (1961). Approximate methods in the quantum theory of many-Fermion systems, *Rev. Mod. Phys.* **33**, 28–36.
[263] Nesbet, R.K. (1961). Construction of symmetry-adapted functions in the many-body problem, *J. Math. Phys.* **2**, 701–709.
[264] Nesbet, R.K. (1968). Analysis of the Harris variational method in scattering theory, *Phys. Rev.* **175**, 134–142.
[265] Nesbet, R.K. (1969). Anomaly-free variational method for inelastic scattering, *Phys. Rev.* **179**, 60–70.
[266] Nesbet, R.K. (1971). Where semiclassical radiation theory fails, *Phys. Rev. Lett.* **27**, 553–556.
[267] Nesbet, R.K. (1979). Energy-modified adiabatic approximation for scattering theory, *Phys. Rev. A* **19**, 551–556.
[268] Nesbet, R.K. (1979). Accurate e^-–He cross sections below 19 eV, *J. Phys. B* **12**, L243–L248.
[269] Nesbet, R.K. (1979). Variational calculations of accurate e^-–He cross sections below 19 eV, *Phys. Rev. A* **20**, 58–70.
[270] Nesbet, R.K. (1980). *Variational Methods in Electron–Atom Scattering Theory* (Plenum, New York).
[271] Nesbet, R.K. (1981). The concept of a local complex potential for nuclear motion in electron–molecule collisions, *Comments At. Mol. Phys.* **11**. 25–35.
[272] Nesbet, R.K. (1984). *R*-matrix formalism for local cells of arbitrary geometry. *Phys. Rev. B* **30**, 4230–4234.
[273] Nesbet, R.K. (1984). Asymptotic distorted-wave approximation for electron–molecule scattering. *J. Phys. B* **17**, L897–L900.
[274] Nesbet, R.K. (1986). Linearized atomic-cell orbital method for energy-band calculations, *Phys. Rev. B* **33**, 8027–8034.
[275] Nesbet, R.K. (1986). Nonperturbative theory of exchange and correlation in one-electron quasiparticle states, *Phys. Rev. B* **34**, 1526–1538.
[276] Nesbet, R.K. (1988). Variational methods for cellular models, *Phys. Rev. A* **38**, 4955–4960.
[277] Nesbet, R.K. (1990). Full-potential multiple scattering theory, *Phys. Rev. B* **41**, 4948–4952.

[278] Nesbet, R.K. (1990). Atomic cell method for total energy calculations, *Bull. Am. Phys. Soc.* **35**, 418.

[279] Nesbet, R.K. (1992). Variational principles for full-potential multiple scattering theory, *Mat. Res. Symp. Proc.* **253**, 153–158.

[280] Nesbet, R.K. (1992). Internal sums in full-potential multiple scattering theory, *Phys. Rev. B* **45**, 11491–11495.

[281] Nesbet, R.K. (1992). Full-potential revision of coherent-potential-approximation alloy theory, *Phys. Rev. B* **45**, 13234–13238.

[282] Nesbet, R.K. (1992). Full-potential multiple scattering theory without structure constants, *Phys. Rev. B* **46**, 9935–9939.

[283] Nesbet, R.K. (1996). Alternative density functional theory for atoms and molecules, *J. Phys. B* **29**, L173–L179.

[284] Nesbet, R.K. (1996). Nonadiabatic phase-matrix method for vibrational excitation and dissociative attachment in electron–molecule scattering, *Phys. Rev. A* **54**, 2899–2905.

[285] Nesbet, R.K. (1997). Fractional occupation numbers in density-functional theory, *Phys. Rev. A* **56**. 2665–2669.

[286] Nesbet, R.K. (1997). Local response model of the generalized polarization potential, *Phys. Rev. A* **56**. 2778–2783.

[287] Nesbet, R.K. (1997). Recent developments in multiple scattering theory and density functional theory for molecules and solids, in *Conceptual Perspectives in Quantum Chemistry*, eds. J.-L. Calais and E. Kryachko (Kluwer, Dordrecht), pp. 1–58.

[288] Nesbet, R.K. (1998). Kinetic energy in density-functional theory, *Phys. Rev. A* **58**, R12–R15.

[289] Nesbet, R.K. (1999). Exact exchange in linear-response theory, *Phys. Rev. A* **60**, R3343–R3346.

[290] Nesbet, R.K. (2000). Bound-free correlation in electron scattering by atoms and molecules, *Phys. Rev. A* **62**, 040701(R).

[291] Nesbet, R.K. (2001). Local potentials in independent-electron models, *Int. J. Quantum Chem.* **81**, 384–388.

[292] Nesbet, R.K. (2002). Orbital functional theory of linear response and excitation, *Int. J. Quantum Chem.* **86**, 342–346.

[293] Nesbet, R.K. and Colle, R. (1999). Does an exact local exchange potential exist? *J. Math. Chem.* **26**, 233–242.

[294] Nesbet, R.K. and Colle, R. (2000). Tests of the locality of exact Kohn–Sham exchange potentials, *Phys. Rev. A* **61**, 012503.

[295] Nesbet, R.K. and Grimm-Bosbach, T. (1993). Use of a discrete complete set of vibrational basis functions in nonadiabatic theories of electron–molecule scattering, *J. Phys. B* **26**, L423–L426.

[296] Nesbet, R.K. and Sun, T. (1987). Self-consistent calculations using canonical scaling in the linearized atomic-cell orbital method: Energy bands of fcc Cu, *Phys. Rev. B* **36**, 6351–6355.

[297] Nesbet, R.K., Mazevet, S. and Morrison, M.A. (2001). Procedure for correcting variational *R*-matrix calculations for polarization response, *Phys. Rev. A* **64**, 034702.

[298] Nesbet, R.K., Noble, C.J. and Morgan, L.A. (1986). Calculations of elastic electron scattering by H_2 for fixed nuclei, *Phys. Rev. A* **34**, 2798–2807.

[299] Nesbet, R.K., Noble, C.J., Morgan, L.A. and Weatherford, C.A. (1984). Variational *R*-matrix calculations of $e^- + H_2$ scattering using numerical asymptotic basis functions, *J. Phys. B* **17**, L891–L895.

[300] Nestmann, B.M. and Peyerimhoff, S.D. (1990). Optimized Gaussian basis sets for representation of continuum wave functions, *J. Phys. B* **22**, L773–L777.
[301] Nestmann, B.M., Nesbet, R.K. and Peyerimhoff, S.D. (1991). A concept for improving the efficiency of R-matrix calculations for electron–molecule scattering, *J. Phys. B* **24**, 5133–5149.
[302] Newton, R.G. (1966). *Scattering Theory of Waves and Particles* (Plenum, New York).
[303] Noble, C.J. and Nesbet, R.K. (1984). CFASYM, a program for the calculation of the asymptotic solutions of the coupled equations of electron collision theory, *Comput. Phys. Commun.* **33**, 399–411.
[304] Noether, E. (1918). *Nachr. Ges. Wiss. Göttingen*, 235.
[305] Norcross, D.W. and Padial, N.T. (1982). The multipole-extracted adiabatic nuclei approximation for electron–molecule collisions, *Phys. Rev. A* **25**, 226–238.
[306] Nordholm, S. and Bacskay, G. (1978). Generalizations of the finite-element and R-matrix methods, *J. Phys. B* **11**, 193–207.
[307] Norman, M.R. and Koelling, D.D. (1984). Towards a Kohn–Sham potential via the optimized effective potential method, *Phys. Rev. B* **30**, 5530–5540.
[308] Nozières, P. (1964). Theory of Fermi liquids, in *1962 Cargèse Lectures in Theoretical Physics*, ed. M. Lévy (Benjamin, New York), IV, 1–37,
[309] Nozières, P. (1964). *Theory of Interacting Fermi Systems* (Benjamin, New York).
[310] Oberoi, R.S. and Nesbet, R.K. (1973). Variational formulation of the R-matrix method for multichannel scattering, *Phys. Rev. A* **8**, 215–219.
[311] Oberoi, R.S. and Nesbet, R.K. (1973). Numerical asymptotic functions in variational scattering theory, *J. Comput. Phys.* **12**, 526–533.
[312] Oberoi, R.S. and Nesbet, R.K. (1974). Addendum to "Variational formulation of the R matrix method for multichannel scattering", *Phys. Rev. A* **9**, 2804–2805.
[313] Oksyuk, Yu.D. (1966). Excitation of the rotational levels of diatomic molecules by electron impact in the adiabatic approximation, *Zh. Eksp. Teor. Fiz.* **49**, 1261–1273. [*Sov. Phys. JETP* **22**, 873–881 (1966)]
[314] O'Malley, T.F. (1966). Theory of dissociative attachment, *Phys. Rev.* **150**, 14–29.
[315] O'Malley, T.F. (1967). Theory of dissociative attachment (Erratum), *Phys. Rev.* **156**, 230.
[316] Ortiz, J.V. (1999). Toward an exact one-electron picture of chemical bonding, *Adv. Quantum Chem.* **35**, 33–52.
[317] Padial, N.T. and Norcross, D.W. (1984). Ro-vibrational excitation of HCl by electron impact, *Phys. Rev. A* **29**, 1590–1593.
[318] Panofsky, W.K.H. and Phillips, M. (1969). *Classical Electricity and Magnetism* (Addison-Wesley, Reading, Mass.).
[319] Parr, R.G. (1983). Density functional theory, *Ann. Rev. Phys. Chem.* **34**, 631–656.
[320] Parr, R.G. and Bartolotti, L.J. (1983). Some remarks on the density functional theory of few-electron systems, *J. Phys. Chem.* **87**, 2810–2815.
[321] Parr, R.G. and Yang, W. (1989). *Density-Functional Theory of Atoms and Molecules* (Oxford University Press, New York).
[322] Pars, L.A. (1962). *An Introduction to the Calculus of Variations* (Wiley, New York).
[323] Pauli, W. (1958). *Theory of Relativity*, tr. G. Field (Pergamon Press, New York).
[324] Payne, P.W. (1979). Density functionals in unrestricted Hartree–Fock theory, *J. Chem. Phys.* **71**, 490–496.
[325] Percival, I.C. and Seaton, M.J. (1957). The partial wave theory of electron-hydrogen atom collisions, *Proc. Camb. Phil. Soc.* **53**, 654–662.

[326] Perdew, J.P. and Norman, M.R. (1982). Electron removal energies in Kohn–Sham density-functional theory, *Phys. Rev. B* **26**, 5445–5450.

[327] Perdew, J.P. and Norman, M.R. (1984). Reply to Comment on 'Electron removal energies in Kohn–Sham density-functional theory', *Phys. Rev. B* **30**, 3525–3526.

[328] Perdew, J.P., Parr, R.G., Levy, M. and Balduz, J.L. (1982). Density-functional theory for fractional particle number: derivative discontinuities of the energy, *Phys. Rev. Lett.* **49**, 1691–1694.

[329] Petersilka, M., Gossmann, U.J. and Gross, E.K.U. (1996). Excitation energies from time-dependent density-functional theory, *Phys. Rev. Lett.* **76**, 1212–1215.

[330] Pfingst, K., Nestmann, B.M. and Peyerimhoff, S.D. (1995). Tailoring the *R*-matrix approach for application to polyatomic molecules, in *Computational Methods for Electron–Molecule Collisions*, eds. W.M. Huo and F. Gianturco (Plenum, New York), pp. 293–308.

[331] Pokorski, S. (1987) *Gauge Field Theories* (Cambridge University Press, Cambridge).

[332] Pulay, P. (1987). Analytical derivative methods in quantum chemistry, *Adv. Chem. Phys.* **69**, 241–286.

[333] Pulay, P. and Fogarasi, G. (1992). Geometry optimization in redundant internal coordinates, *J. Chem. Phys.* **96**, 2857–2860.

[334] Purcell, J.E. (1969). Nuclear structure calculations in the continuum-application to neutron-carbon scattering, *Phys. Rev.* **185**, 1279–1285.

[335] Quigg, C. (1983). *Gauge Theories of the Strong, Weak, and Electromagnetic Interactions* (Benjamin-Cummings, New York).

[336] Ramond, P. (1989). *Field Theory: A Modern Primer*, 2nd edition (Addison-Wesley, Redwood City, California).

[337] Lord Rayleigh (1892). *Phil. Mag.* **34**, 481.

[338] Lord Rayleigh (1892). *The Theory of Sound*, 2nd edition, Vol. 2, p. 328, reprinted by Dover Publications, New York (1976).

[339] Reed, M. and Simon, B. (1980). *Methods of Modern Mathematical Physics I: Functional Analysis* (Academic Press, New York).

[340] Renton, P. (1990). *Electroweak Interactions* (Cambridge University Press, New York).

[341] Rescigno, T.N., McCurdy, C.W., Orel, A.E. and Lengsfield, B.H., III (1995). The complex Kohn variational method, in *Computational Methods for Electron–Molecule Collisions*, eds. W.M. Huo and F.A. Gianturco (Plenum, New York), pp. 1–44.

[342] Rescigno, T.N., McCurdy, C.W. and McKoy, V. (1974). Discrete basis set approach to nonspherical scattering, *Chem. Phys. Lett.* **27**, 401–404.

[343] Rescigno, T.N., McCurdy, C.W. and McKoy, V. (1974). Discrete basis set approach to nonspherical scattering. II., *Phys. Rev. A* **10**, 2240–2245.

[344] Rescigno, T.N., McCurdy, C.W. and McKoy, V. (1975). Low energy e^-–H_2 elastic cross sections using discrete basis functions, *Phys. Rev. A* **11**, 825–829.

[345] Rivers, R.J. (1987). *Path Integral Methods in Quantum Field Theory* (Cambridge University Press, New York).

[346] Rohr, K. and Linder, F. (1976). Vibrational excitation of polar molecules by electron impact I. Threshold resonances in HF and HCl, *J. Phys. B* **9**, 2521–2537.

[347] Roothaan, C.C.J. (1951). New developments in molecular orbital theory, *Rev. Mod. Phys.* **23**, 69–89.

[348] Roothaan, C.C.J. (1960). Self-consistent field theory for open shells of electronic systems, *Rev. Mod. Phys.* **32**, 179–185.

[349] Roothaan, C.C.J. and Bagus, P.S. (1963). Atomic self-consistent field calculations by the expansion method, *Methods Comput. Phys.* **2**, 47–94.

[350] Rubinow, S.I. (1955). Variational principle for scattering with tensor forces, *Phys. Rev.* **98**, 183–187.

[351] Runge, E. and Gross, E.K.U. (1984). Density-functional theory for time-dependent systems, *Phys. Rev. Lett.* **52**, 997–1000.

[352] Ryder, L.H. (1985). *Quantum Field Theory* (Cambridge University Press, Cambridge).

[353] Salam, A. (1968). In *Proceedings of the Eighth Nobel Symposium*, ed. N. Svartholm (Almqvist and Wiksell, Stockholm).

[354] Sarpal, B.J., Tennyson, J. and Morgan, L.A. (1991). Vibrationally resolved electron HeH^+ collisions using the nonadiabatic R-matrix method, *J. Phys. B* **24**, 1851–1866.

[355] Sarpal, B.J., Tennyson, J. and Morgan, L.A. (1994). Dissociative recombination without curve crossing: study of HeH^+, *J. Phys. B* **27**, 5943–5953.

[356] Schlegel, H.B. (1982). Optimization of equilibrium geometries and transition structures, *J. Comput. Chem.* **3**, 214–218.

[357] Schlegel, H.B. (1987). Optimization of equilibrium geometries and transition structures, *Adv. Chem. Phys.* **67**, 249–286.

[358] Schlessinger, L. and Payne, G.L. (1974). Procedure for finding the scattering solutions of the Schrödinger equation, *Phys. Rev. A* **10**, 1559–1567.

[359] Schlosser, H. and Marcus, P. (1963). Composite wave variational method for solution of the energy-band problem in solids, *Phys. Rev.* **131**, 2529–2546.

[360] Schneider, B.I. (1975). R-matrix theory for electron–atom and electron–molecule collisions using analytic basis set expansions. *Chem. Phys. Lett.* **31**, 237–241.

[361] Schneider, B.I. (1976). Role of the Born–Oppenheimer approximation in the vibrational excitation of molecules by electrons, *Phys. Rev. A* **14**, 1923–1925.

[362] Schneider, B.I. (1995). An R-matrix approach to electron–molecule collisions, in *Computational Methods for Electron–Molecule Collisions*, eds. W.M. Huo and F. Gianturco (Plenum, New York), pp. 213–226.

[363] Schneider, B.I., LeDourneuf, M. and Burke, P.G. (1979). Theory of vibrational excitation and dissociative attachment: an R-matrix approach, *J. Phys. B* **12**, L365–L369.

[364] Schneider, B.I., LeDourneuf, M. and Vo Ky Lan (1979). Resonant vibrational excitation of N_2 by low-energy electrons: an *ab initio* R-matrix calculation, *Phys. Rev. Lett.* **43**, 1926–1929.

[365] Schrödinger, E. (1926). Quantisiering als Eigenwertproblem, *Ann. der Physik* **81**, 109–139.

[366] Schulman, L.S. (1981). *Techniques and Applications of Path Integration* (Wiley, New York).

[367] Schulz, G.J. (1962). Vibrational excitation of nitrogen by electron impact, *Phys. Rev.* **125**, 229–232.

[368] Schulz, G.J. (1964). Vibrational excitation of N_2, CO, and H_2 by electron impact, *Phys. Rev.* **135**, A988–A994.

[369] Schwartz, C. (1961). Variational calculations of scattering, *Ann. Phys. (N.Y.)* **16**, 36–50.

[370] Schwartz, C. (1961). Electron scattering from hydrogen, *Phys. Rev.* **124**, 1468–1471.

[371] Schwinger, J. (1947). A variational principle for scattering problems, *Phys. Rev.* **72**, 742.

[372] Schwinger, J. (1951). On the Green's functions of quantized fields, *Proc. Nat. Acad. Sci.* **37**, 452–455.

[373] Schwinger, J. (1958). *Quantum Electrodynamics* (Dover, New York).

[374] Seaton, M.J. (1953). The Hartree–Fock equations for continuous states with applications to electron excitation of the ground configuration terms of O I, *Phil. Trans. Roy. Soc. (London) A* **245**, 469–499.

[375] Seaton, M.J. (1966). A variational principle for the scattering phase matrix, *Proc. Phys. Soc. (London)* **89**, 469–470.

[376] Seaton, M.J. (1973). Close coupling, *Comput. Phys. Commun.* **6**, 247–256.

[377] Seaton, M.J. (1983). Quantum defect theory, *Rep. Prog. Phys.* **46**, 167–257.

[378] Seitz, L.I. (1955). *Quantum Mechanics*, 2nd edition (McGraw-Hill, New York).

[379] Shanno, D.F. (1970). Conditioning of quasi-Newton methods for function minimisation, *Math. Comput.* **24**, 647–656.

[380] Sharp, R.T. and Horton, G.K. (1953). A variational approach to the unipotential many-electron problem, *Phys. Rev.* **90**, 317.

[381] Shimamura, I. (1977). In *The Physics of Electronic and Atomic Collisions*, invited papers, X ICPEAC, Paris (1977), (North-Holland, Amsterdam).

[382] Siegert, A.J.F. (1939). On the derivation of the dispersion formula for nuclear reactions, *Phys. Rev.* **56**, 750–752.

[383] Sinfailam, A.L. and Nesbet, R.K. (1972). Variational calculations of electron–helium scattering, *Phys. Rev. A* **6**, 2118–2125.

[384] Skriver, H.L. (1984). *The LMTO Method* (Springer-Verlag, Berlin).

[385] Slater, J.C. (1951). A simplification of the Hartree–Fock method, *Phys. Rev.* **81**, 385–390.

[386] Slater, J.C. (1972). Statistical exchange-correlation in the self-consistent field, *Adv. Quantum Chem.* **6**, 1–92.

[387] Slater, J.C. (1974). *The Self-Consistent Field for Molecules and Solids* (McGraw-Hill, New York).

[388] Slater, J.C. and Wood, J.H. (1971). Statistical exchange and the total energy of a crystal, *Int. J. Quantum Chem. Suppl.* **4**, 3–34.

[389] Smith, F.T. (1960). Lifetime matrix in collision theory, *Phys. Rev.* **118**, 349–356.

[390] Sundermeyer, K. (1983). *Constrained Dynamics* (Springer-Verlag, Berlin).

[391] Swanson, M.S. (1992). *Path Integrals and Quantum Processes* (Academic Press, New York).

[392] Symanzik, K. (1954). Schwinger's functional in field theory, *Z. Natürforschung* **9a**, 809–824.

[393] Synge, J.L. (1956). *Relativity: the Special Theory* (Interscience, New York).

[394] Szabo, A. and Ostlund, N. S. (1982). *Modern Quantum Chemistry; Introduction to Advanced Electronic Structure Theory* (McGraw-Hill, New York).

[395] Takahashi, Y. (1957). On the generalized Ward identity, *Nuovo Cimento* **6**, 371–375.

[396] Takatsuka, K. and McKoy, V. (1981). Extension of the Schwinger variational principle beyond the static exchange approximation, *Phys. Rev. A* **24**, 2473–2480.

[397] Takatsuka, K., Lucchese, R.R. and McKoy, V. (1981). Relationship between the Schwinger and Kohn-type variational principles in scattering theory, *Phys. Rev. A* **24**, 1812–1816.

[398] Talman, J.D. and Shadwick, W.F. (1976). Optimized effective atomic central potential, *Phys. Rev. A* **14**, 36–40.

[399] Taylor, J.T. (1983). *Scattering Theory* (Krieger, FL).

[400] Teller, E. (1962). On the stability of molecules in the Thomas–Fermi theory, *Rev. Mod. Phys.* **34**, 627–631.

[401] Temkin, A. (1976). Some recent developments in electron–molecule scattering: theory and calculation, *Comments At. Mol. Phys.* **5**, 129–139.
[402] Thomas, L.H. (1927). The calculation of atomic fields, *Proc. Camb. Phil. Soc.* **23**, 542–548.
[403] Thompson, C.J. (1972). *Mathematical Statistical Mechanics* (Princeton University Press, Princeton, New Jersey).
[404] 't Hooft, G. (1971). Renormalization of massless Yang–Mills fields, *Nucl. Phys.* **B33**, 173–199.
[405] 't Hooft, G. (1971). Renormalizable Lagrangians for massive Yang–Mills fields, *Nucl. Phys.* **B35**, 167–188.
[406] 't Hooft, G. and Veltman, M. (1972). Regularization and renormalization of gauge fields, *Nucl. Phys.* **B44**, 189–213.
[407] Thouless, D.J. (1961). *The Quantum Mechanics of Many-body Systems* (Academic Press, New York).
[408] Thümmel, H.T., Grimm-Bosbach, T., Nesbet, R.K. and Peyerimhoff, S.D. (1995). Rovibrational excitation by electron impact, in *Computational Methods for Electron–Molecule Collisions*, eds. W.M. Huo and F. Gianturco (Plenum, New York), pp. 265–291.
[409] Thümmel, H.T., Peyerimhoff, S.D. and Nesbet, R.K. (1992). Near-threshold rotational excitation in electron–polar molecule scattering, *J. Phys. B* **25**, 4553–4579.
[410] Thümmel, H.T., Peyerimhoff, S.D. and Nesbet, R.K. (1993). Near-threshold rovibrational excitation of HF by electron impact, *J. Phys. B* **26**, 1223–1251.
[411] Trail, W.K., Morrison, M.A., Isaacs, W.A. and Saha, B.C. (1990). Simple procedures for including vibrational effects in the calculation of electron–molecule cross sections, *Phys. Rev. A* **41**, 4868–4878.
[412] Trickey, S.B. (1984). Comment on 'Electron removal energies in Kohn–Sham density-functional theory', *Phys. Rev. B* **30**, 3523.
[413] Volterra, V. (1954). *Opere Matematiche*, 5 vols. (Accademia dei Lincei, Rome, 1954–1962).
[414] von Barth, U. and Hedin L. (1972). A local exchange-correlation potential for the spin-polarized case: I, *J. Phys. C* **5**, 1629–1642.
[415] Wadehra, J.M. (1984). Dissociative attachment to rovibrationally excited H_2, *Phys. Rev. A* **29**, 106–110.
[416] Wadehra, J.M. and Bardsley, J.N. (1978). Vibrational and rotational state dependence of dissociative attachment in $e-H_2$ collisions, *Phys. Rev. Lett.* **41**, 1795–1798.
[417] Ward, J.C. (1950). The scattering of light by light, *Phys. Rev.* **77**, 293.
[418] Watson, D.K. (1988). Schwinger variational methods, *Adv. At. Mol. Phys.* **25**, 221–250.
[419] Watson, D.K. and McKoy, V. (1979). Discrete basis function approach to electron–molecule scattering, *Phys. Rev. A* **20**, 1474–1483.
[420] Weinberg, S. (1967). A model of leptons, *Phys. Rev. Lett.* **19**, 1264–1266.
[421] Weinberg, S. (1995). *The Quantum Theory of Fields*, Vol. I (Cambridge University Press, New York).
[422] Weinberg, S. (1996). *The Quantum Theory of Fields*, Vol. II (Cambridge University Press, New York).
[423] Weiner, B. and Trickey, S.B. (1999). Time-dependent variational principle in density functional theory, *Adv. Quantum Chem.* **35**, 217–247.
[424] Weyl, H. (1929). *Z. Phys.* **56**, 330.
[425] Weyl, H. (1951). *Space, Time, Matter* (Dover, New York).

[426] Wheeler, J.A. and Feynman, R.P. (1945). Interaction with the absorber as the mechanism of radiation, *Rev. Mod. Phys.* **17**, 157–179.

[427] Wigner, E.P. (1948). On the behavior of cross sections near thresholds, *Phys. Rev.* **73**, 1002–1009.

[428] Wigner, E.P. and Eisenbud, L. (1947). Higher angular momenta and long range interaction in resonance reactions, *Phys. Rev.* **72**, 29–41.

[429] Wigner, E.P. and Seitz, F. (1933). On the constitution of metallic sodium, *Phys. Rev.* **43**, 804–810.

[430] Williams, A.R., Kübler, K. and Gelatt, C.D. (1979). Cohesive properties of metallic compounds: Augmented-spherical-wave calculations, *Phys. Rev. B* **19**, 6094–6118.

[431] Williams, A.R. and Morgan, J. van W. (1972). Multiple scattering by non-muffin-tin potentials, *J. Phys. C* **5**, L293–L298.

[432] Williams, A.R. and Morgan, J. van W. (1974). Multiple scattering by non-muffin-tin potentials: general formulation, *J. Phys. C* **7**, 37–60.

[433] Wolfenstein, L. (1999). Neutrino physics, *Rev. Mod. Phys.* **71**, S140–S144.

[434] Yang, C.N. (1977). Magnetic monopoles, fiber bundles, and gauge fields, *Ann. N.Y. Acad. Sci.* **294**, 86–97.

[435] Yang, C.N. and Mills, R.L. (1954). Conservation of isotopic spin and isotopic gauge invariance, *Phys. Rev.* **96**, 191–195.

[436] Yourgrau, W. and Mandelstam, S. (1968). *Variational Principles in Dynamics and Quantum Theory*, 3rd edition (Dover, New York).

[437] Zeller, R. (1987). Multiple-scattering solution of Schrödinger's equation for potentials of general shape, *J. Phys. C* **20**, 2347–2360.

[438] Zeller, R. (1988). Empty-lattice test for non-muffin-tin multiple-scattering equations, *Phys. Rev. B* **38**, 5993–6002.

[439] Zerner, M.C. (1989). Analytic derivative methods and geometry optimization, in *Modern Quantum Chemistry, Introduction to Advanced Electronic Structure Theory*, eds. A. Szabo and N.S. Ostlund (McGraw-Hill, New York), pp. 437–458.

[440] Zhang, J.Z.H. and Miller, W.H. (1989). Quantum reactive scattering via the S-matrix version of the Kohn variational principle: Differential and integral cross sections for $D + H_2 \to HD + H$, *J. Chem. Phys.* **91**, 1528–1547.

[441] Zhang, X.-G. and Butler, W.H. (1992). Simple cellular method for the exact solution of the one-electron Schrödinger equation, *Phys. Rev. Lett.* **68**, 3753–3756.

[442] Zhang, X.-G. and Butler, W.H. (1992). Multiple-scattering theory with a truncated basis set, *Phys. Rev. B* **46**, 7433–7447.

[443] Zhang, X.-G., Butler, W.H., MacLaren, J.M. and van Ek, J. (1994). Cellular solutions for the Poisson equation in extended systems, *Phys. Rev. B* **49**, 13383–13393.

[444] Zhang, X.-G., Butler, W.H., Nicholson, D.M. and Nesbet, R.K. (1992). Green-function cellular method for the electronic structure of molecules and solids, *Phys. Rev. B* **46**, 15031–15039.

[445] Zhao, Q. and Parr, R.G. (1993). Constrained-search method to determine wave functions from electronic densities, *J. Chem. Phys.* **98**, 543–548.

[446] Zhao, Q., Morrison, R.C. and Parr, R.G. (1994). From electron densities to Kohn–Sham kinetic energies, orbital energies, exchange-correlation potentials, and exchange-correlation energies, *Phys. Rev. A* **50**, 2138–2142.

[447] Zumbach, G. and Maschke, K. (1983). New approach to the calculation of density functionals, *Phys. Rev. A* **28**, 544–554.

[448] Zvijac, D.J., Heller, E.J. and Light, J.C. (1975). Variational correction to Wigner R-matrix theory of scattering, *J. Phys. B* **8**, 1016–1033.

[426] Wheeler J.A. and Feynman R.P. (1945) Interaction with the absorber as the mechanism of radiation, *Rev. Mod. Phys.* **17**, 157–1[illegible]
[427] Wigner E.P. (1948) On the behavior of cross sections near thresholds, *Phys. Rev.* **73**, 1002–1009
[428] Wilner [illegible] and Fischbach [illegible] (19[illegible]) [illegible] and [illegible] interaction [illegible] reactions, *Phys. Rev.* **72**, [illegible]
[429] W[illegible] (1949) [illegible], *Rev.* [illegible]
[illegible] 808–810
[430] Williams A.F. [illegible] K. and [illegible] C.D. (19[illegible]) [illegible] properties of [illegible] compounds: Augmented spherical [illegible] calculations, *Phys. Rev.* **B** [illegible]
6004–6[illegible]
[431] Williams A. [illegible] (19[illegible]) [illegible] potentials, *J. Phys.* [illegible]
[432] Williams A. [illegible] and [illegible] (19[illegible]) [illegible] in [illegible], *Phys.* [illegible]–661
[illegible]

Index